12 Good Reasons to Look Up Uranus

12 Good Reasons to Look Up Uranus

Kevin Joslin

Legend Press Ltd, 2 London Wall Buildings,
London EC2M 5UU
info@legend-paperbooks.co.uk
www.legendpress.co.uk

British Library Cataloguing in Publication Data available.

ISBN 978-1-9077563-6-8

All characters, other than those clearly in the public domain, and place names, other than those well-established such as towns and cities, are fictitious and any resemblance is purely coincidental.

Set in Times

Printed by JF Print Ltd., Sparkford.

Cover designed by Gudrun Jobst
www.yotedesign.com

ALSO BY KEVIN JOSLIN:

See John Run

The Complete Radio 2

Janet & John Marsh Stories

As told by Terry Wogan

www.togs.org

For Lucy

A bright light in a dull world

Foreword

Gypsy Petulengro, Mystic Meg, Prince Monolulu, My Aunt Nellie with the tea-leaves. All of them practised in the dark arts; in the great tradition of soothsayers and necromancers...

Anyone who deludes themselves that Kevin Joslin is part of this tradition can only be regarded as an escapee from the Home for the Bewildered. Don't let the earrings and crystal ball fool you; this man is the son of a disgraced bookies runner from Cahirciveen. Not only does he know nothing of the future, he has little grasp of the present. Do not cross this numpty's palm with silver!

Terry Wogan

INTRODUCTION

Ever since primitive man first gazed up into the eternity of night and asked himself the question, 'What is that stuff I just stepped in?' people have wondered what effect the star-strewn heavens have on their daily lives.

Following years of painstaking research, this volume of precise and detailed astrological forecasts for the year ahead, eschews the normal wishy-washy and rather all encompassing language of the hack astrologers who typically ply their trade in daily newspapers, the editors of which wouldn't know a real astrologer even if they had put up a tent on their front lawn and played Ink Spots records at deafening volume on a ghetto-blaster until they agreed to see the person in question, who incidentally has no connection to myself or anyone I know.

No. We deal with specifics here. If your cat is about to be flattened by a falling space-station component, you will know about it in advance, and can take immediate action to avoid the impending disaster – assuming that you actually like your cat that is.

The week ahead for each of the twelve signs of the zodiac is forecast in summary form, with specific events highlighted along with any items that are particularly imbued with good fortune.

Should any of the forecasts prove at all inaccurate, you may of course return the book to the retailer, who will, I confidently predict, roll their eyes and say to a colleague, 'They're out in force this morning Jeff'.

KEVIN JOSLIN

Weekly Forecast for 3rd to 9th January

Aries

The Moon joins Venus, the planet of relationships, in the area of your chart that governs income tax. You may have to adopt an extremely unconventional approach with the tax inspector to reduce your next bill. Saturn rising means that it would be best to try out those new soup spoons by next Friday, when Mars goes trine.

Lucky cake: Caraway seed
Lucky bag: Gladstone

Taurus

Before Mercury goes retrograde next week, it would be a good idea to spend as much time as possible away from the office. Charcoal insoles are likely to gain new significance. On Friday, you will be surprised when your passion for knitting is discovered when you are caught with your nose in a *Woman's Realm* in the public library.

Lucky sport: Jousting
Lucky vault: Double-twisting Tsukahara

Gemini

After next Friday things may not run quite so smoothly for a few weeks, while your ruler, Mercury, is retrograde. Tuesday's New Moon will see you in the midst of things socially, but a misheard comment about Marigold gloves and Swarfega will lead to an unseemly scuffle at the Bingo on Wednesday. You may live to regret having given up the piano-accordion.

Lucky manoeuvre: Heimlich
Lucky Walton: Jim-Bob

Cancer

The Moon joins Venus, the planet of relationships, in the area of your chart that governs junk-food means that your partner makes the shocking discovery that for the last three and a half years you have been doing something unusual with the malted-milk biscuits. Instead of eating them conventionally, you have been nibbling around the edges until only the cow-shaped portion of biscuit remains, and hiding them in your sock drawer then playing with the herd on the dressing table.

Lucky stain: Mustard
Lucky pancake: Scotch

Leo

It's a busy time. You may start the week in a whirl of activity, but on the whole it is positive and productive until Tuesday lunchtime when you fall into bad company and wake up dressed only in Tupperware and tied to a telegraph pole in Low Moor Road, Bradford.

Lucky colour: Periwinkle
Lucky vitamin: Riboflavin

Virgo

Your ruler, Mercury, goes retrograde next Friday, which means an unsettled period ahead. It might seem a hard thing to achieve, but you must get over your first love. No good will come of those conjugal visits in prison. Learn to enjoy the lascivious stares of the Finance Director in the queue at the whelk stall. There's a lot more to him than meets the eye.

Lucky eye shadow: Blue
Lucky pen: Ballpoint

Libra

Tuesday's New Moon is in dispute with Pluto. It's likely to go to arbitration but will almost certainly be resolved before it gets to a formal hearing. The Sun in Capricorn means that there is a bit of an atmosphere at work. On the way to the office you witness a UFO abduction by the railway station, but you have other things on your mind, so forget to mention it.

Lucky snack: Quail eggs
Lucky Stone: Bill Wyman

Scorpio

Well, you're a darkhorse and that's for sure! Not only have you managed to conceal the fact from your friends that you've had a big win on the bingo, a benign Saturn reveals that you have recently patented a method for turning split-ends into weapons-grade plutonium. Mind how you cross the road on Wednesday.

Lucky seven: Samurai
Lucky soft drink: Um-Bongo

Sagittarius

Mars in your sign is likely to cause domestic upheaval this week. Expect an argument on Tuesday involving a nine-iron and a meat and potato pie. A quiet man in white gloves will take on new significance in your life on Wednesday. The New Moon in Capricorn will bring welcome respite from that mysterious whooping noise from your back boiler.

Lucky lunch: Chops and mash
Lucky trousers: Corduroy

Capricorn

Mercury rising means you would do well to avoid men with 'Blakeys' in their shoes this week. Your normal good humour will be sorely tested by an irascible colleague on Thursday. If you are prone to in-growing toenails, you should avoid Warwick Castle at all costs.

Lucky colour: Heliotrope
Lucky landseer: Monarch of the Glen

Aquarius

As Tuesday's New Moon approaches, a close friend may bring you good news about your dandruff. There can only be one outcome should you choose to ignore the frayed elastic in your ill-starred foundation garments. If you have Venus rising at the end of the week, you only have yourself to blame. However, there's a good chance of new trousers on Saturday.

Lucky red: Barolo
Lucky gland: Pancreas

Pisces

As the period of change that began with the New Moon moves into a new stage, you will be less troubled by persistent verucca than in recent months although hard skin will still prove problematic. Don't be tempted by a bargain fireguard until Mercury goes retrograde on Friday.

Lucky weakness: Turkish Delight
Lucky dance: The Watutsi

Weekly Forecast for 10th to 17th January

Aries

This week you will be in your element (Helium). Neptune indicates that surprises are in store towards the end of the week when you may be involved in an incident with a Glockenspiel – and it's no use thinking that broad beans will solve your problems this time.

Lucky vessel: Schooner
Lucky cartoon: Squiddly-Diddly

Taurus

The New Moon last week indicates that now is the time to clean out that cupboard under the stairs. While doing so, you find a secret passage to Lerwick in the Shetland Islands. When Mercury goes retrograde, you will receive an unexpected invitation to afternoon tea at the Ritz with Shabba Ranks.

Lucky number: 2,124
Lucky pants: Blue Y-Fronts

Gemini

The Sun, your ruler, in your opposite sign of Aquarius, is trine Saturn, which indicates that you should make sure you don't run out of lip-balm on Wednesday. If you see an unusually shaped vegetable, buy it; it will see off an armed intruder.

Lucky snack: Toast Toppers
Lucky game: Connect 4

Cancer

You would be well advised to pick up a new bird-bath on Tuesday as the Sun forming a trine to Saturn, emphasised by the New Moon, means that you will encounter some really dirty birds later in the week. Although your ruler Mercury is retrograde, even your dodgy instep feels better.

Lucky Turner: *The Fighting Temeraire*
Lucky smell: Creosote

Leo

The early part of the week may bring you an unexpected message from the past – perhaps an old flame, or possibly the results of a blood test. Tuesday lunchtime will see you cornered by a stocky Yorkshireman wearing a tie that could best be described as 'a bit sudden'. Under no circumstances should you accompany him to the Post Office.

Lucky lubricant: WD40
Lucky marmalade: Grapefruit

Virgo

After a weak Saturn/Uranus, Saturday will bring a new passion into your life when you come across a hitherto undiscovered aptitude for Scottish country dancing. Sagittarius rising will wake you in the early hours of Sunday by slamming the car door. The weekend will bring some respite as you discover a new floor-cleaning product while attending a baptism.

Lucky number: Pi
Lucky prophet: Isiah

Libra

On Thursday you must decide whether to turn left or right, one road leads to sorrow whilst the other leads to a full and happy life. There are no signs to help you on your way but the road to the left has a decent pub on it. You would do well to remember that he who lives in a glass house shouldn't invite he who is without sin over for drinks.

Lucky force: Centrifugal
Lucky soup: Oxtail

Scorpio

It's a great time for you to do things in a group this week – particularly after your suspicions about the number of cars outside number 52 every weekend are confirmed on Wednesday evening when you pop round to borrow a cup of Swarfega. An unpleasant aspect between Mars and Neptune means that you should probably skip the clams this week.

Lucky biscuit: Gypsy Creams
Lucky ointment: Firey Jack

Sagittarius

You've been a bit worried about a squeaking noise from the front nearside brake on your car over the weekend. Saturn entered the area of your chart concerning wildlife on Saturday, so it will probably turn out to be Finches again. An interesting aspect between Neptune and Saturn indicates that you should accept the long-standing offer of marriage from the local pharmacist.

Lucky hedge: Blackthorn
Lucky nostril: Left

Capricorn

Venus joining forces with Jupiter this week indicates that you will at last fulfil your dream to become a Welsh language square-dance caller and turn your back on your humdrum day-job. On Thursday you are likely to need a new coil fitted. This is best left to the AA.

Lucky pliers: Snipe-nosed
Lucky street party: The M4

Aquarius

An interesting week during which new opportunities present themselves. Maybe part-time job or money-making hobby. Take care not to overstretch yourself as a trine Pluto can often mean that you work long hours for scant reward – just as you do already. Last week's New Moon rising indicates that you will pick up a splinter while working on a Japanese whaling ship at the weekend.

Lucky GCSE: Forestry
Lucky Bishop: Desmond Tutu

Pisces

Even though you may have recently enjoyed a few days off, Mars rising indicates that there are some tensions in your life. Your partner's continued enthusiasm for the rough-and-tumble world of paintballing may be one source. Try to find out who has been plotting against you while you were away. Don't just blame the Pixies this time either.

Lucky topping: Hundreds and Thousands
Lucky roofing: Chestnut shingles

Weekly Forecast for 24th to 30th January

Aries

Stubborn Pluto in your sign means that your refusal to dye your hair blonde like the rest of the administrative staff is earning you no friends among the management this week. On Monday, the Moon in Neptune indicates that a well-nourished, tanned man may try to interest you in looking at his holiday photographs. Try not to snigger.

Lucky lentils: Puy
Lucky rub: Witch-hazel

Taurus

Kindly Jupiter enters your sign on Wednesday removing any remaining doubts you might have had about your partner's plan to take over a whelk-stall franchise in Slough. The end of the week sees you among friends and admirers – perhaps an outing is planned. Everything will go well until Friday lunchtime when you are in Debenhams and are struck by lightning in the trouser department.

Lucky pint: Lager-top
Lucky cad: Stewart Grainger

Gemini

This week, don't allow yourself to be distracted. Usually people with Gemini strong in their charts like to have a bit of this and a bit of the other, keeping their hand in with everybody. On Monday, Mars entering your fourth house means that you could do well in matters involving Lino. Try not to look too smug.

Lucky saga: Beowulf
Lucky joint: Mortice & Tenon

Cancer

After an exhausting weekend, your ruler, the Moon, moves into dynamic Taurus, filling you with unusual impulses. You may make a new friend this week, or find that an old hobby becomes interesting again. You could meet someone from far away, or even turn to a life of crime. Important days will be Wednesday and Thursday. Important buses will be the 37A and the 49B from the High Street.

Lucky restorative: Wincarnis
Lucky move: Knight to Kings-Bishop 5

Leo

However ambitious you are, you like to have your home to return to and then close the door. This week, however, you feel like flinging the door open and offering invitations to all and sundry, your friends, family, and anyone involved in marine-insurance. Watch out for a drama involving blue socks and a vintage threshing machine on Thursday.

Lucky weed: Cleavers
Lucky polish: Mr. Sheen

Virgo

There seems to be conflict in the area of your chart concerned with replacement windows. A trine Pluto will almost certainly mean trouble over setting your neighbour's Aunt on fire on Thursday. A rare breed of Chinchilla and a pint of out-of-date mushroom soup will unexpectedly provide the perfect alibi.

Lucky excuse: Pyrokenesis
Lucky accelerant: Four star

Libra

Saturn goes retrograde in your sign on Tuesday, which indicates that you should think carefully about your entry for the world pro-celebrity Buckaroo championships in Chelmsford. You can take advantage of the vogue for live action remakes of cartoon shows, as your ideal role of 'Foghorn Leghorn' is up for grabs at the end of the week.

Lucky sauce: Worcester
Lucky knot: Sheep-shank

Scorpio

Capricorn rising and a troublesome aspect between the Sun and Pluto means that on Wednesday the wine merchants will mess up your sherry order again, leaving you a bottle of cream for your dry sack. Everybody makes mistakes, so try not to rub it in. The Moon rising in Aquarius indicates that new net-curtains are on the cards.

Lucky snack: Liquorice bootlaces
Lucky fragrance: Old Spice

Sagittarius

Saturn has turned retrograde in your sign this week. After the relaxing few days you've recently enjoyed this will mean a return to the same old routine. But looking on the bright side, that rash that you thought might be Ebola actually turns out to be just Scurvy.

Lucky island: Canvey
Lucky trade: Haberdashery

Capricorn

Your Venus is looking increasingly ill-at-ease of late which may help to explain the trouble you've had getting satisfactory results from your new shoe-trees. On Tuesday, a trine Saturn indicates that the blue corduroys you bought last week will go baggy in the seat on their first wash.

Lucky dumplings: Dim-sum
Lucky alias: Mano-Lito Montoya

Aquarius

A truly momentous week during which you will discover the source of both the rumours about you and Angela Merkel, and the recipe for first class gazpacho. Uranus being square may indicate health worries toward the end of the week. The weekend brings news that the lucrative work you have been receiving as a Gordon Brown look-alike may be coming to an end.

Lucky plant: Robert
Lucky container: Punnet

Pisces

You like to make your decisions and stick to them, but if you are not flexible and open minded this week, all of the plans you've made will fail to bear fruit, or at least not fruit that you enjoy very much. Your partner's continuing enthusiasm for tofu is a bit of a worry.

Lucky attitude: Munificent
Lucky currency: The Belgian Bun

Weekly Forecast for 31st January to 6th February

Aries

At the beginning of the week, take time out with people close to you. Some relationships may be intense, others in huts and outbuildings. Tuesday will be a particularly good day for sorting out that sticky letterbox flap that lets in the draught that your neighbour keeps on about.

Lucky work-surface: Formica
Lucky greens: Pak choi

Taurus

On Tuesday, romantic Venus forms an important trine with Saturn, bringing with it fond memories of your youth tempered with the sad realisation that being spanked every lunchtime by a middle-aged woman who smelled of camphor was not likely to happen again – Particularly not in the saloon bar. Thursday brings a change of fortune when Pork loins are on special in Waitrose.

Lucky canal: Alimentary
Lucky attitude: Raffish

Gemini

On Monday, the sudden revelation that all of these 'so-called hip-hop M.C.s' are doing nothing more than 'mucking about with record players' signals a career change. You discover that people will pay good money to watch you perform on stage with a Sandwich Toaster. Pluto entering your birthsign on Friday could mean a nasty accident with molten cheese.

Lucky involuntary movement: Shudder
Lucky drink: The usual

Cancer

Many people see Cancerians as a little timorous, insecure, clinging to the familiar rather than striking out for what is new. They don't see your large collection of restraints, dungeon equipment, and the advertisements placed in the local press to lure unwary tradesmen. On Thursday Mars moves into your sign, which indicates that the security guard at work will mention your oubliette to a local celebrity.

Lucky jug: Toby
Lucky Primate: The Archbishop of York

Leo

Although your last Birthday Karaoke session was enjoyable, you finally recognise that your dream of making a top 20 single is likely to be prejudiced by the fact that you have ear-hairs older than Lady Gaga. Neptune rising will trigger an allergic reaction to Germolene. Be sure to let Hortense at the sandwich bar know about this.

Lucky smell: Play-Doh
Lucky custard: Banana

Virgo

Jupiter's influence will wane this week, and with it any hopes you might have entertained of escaping the forthcoming shopping trip to Bluewater. On Thursday the Moon transits Pluto, which usually means that someone will step on your bad toe while line dancing.

Lucky lemur: Ring-tailed
Lucky present: Nine Lords a leaping

Libra

The appearance of Chiron transiting the New Moon on Thursday will bring a new and powerful influcnce to bear on your life. You are spotted while leaping to avoid a cycle-courier outside the office and the new support tights you bought last Wednesday inadvertently earn you a place on the Olympic long-jump team.

Lucky soup: Mulligatawny
Lucky finish: Treble 14, double top

Scorpio

Mars entering your birthsign this week signals a change of fortune when England's opening pair score over 237,000 runs off the first seven overs, you discover that you have been selected to Captain the next World Cup team, and you are recognised in the corridor by a member of the Board. See if you can get spread-betting odds on the first two.

Lucky air: Professional detachment
Lucky duck: Bombay

Sagittarius

A square Saturn this week indicates that on Wednesday you may have a spot of trouble when the foot-spa you'd set your heart on is out of stock in Argos. Try to remain calm – rubbing strawberry cheesecake into the manager's hair is unlikely to help matters.

Lucky wallpaper: Woodchip
Lucky topping: Almond flakes

Capricorn

Pluto rising means that this week you finally realise that you're surrounded by vacuous well-wishers and slight acquaintances who deliberately try to steer you away from your destiny to bolster their own flagging careers. You find yourself itching to use the knuckle-dusters the Vicar bought you for Christmas.

Lucky affliction: Nervous tic
Lucky disguise: Rob Brydon

Aquarius

A note of caution this week. Retrograde Saturn means that if you deal with others individually, success will greet you. Deal with them as a group and you will be eaten by escaped Wolverines in Potters Bar. On Thursday, a close relative may forget to post you some clean socks for your forthcoming trial.

Lucky number: Patrick McGoohan
Lucky escape: February 5th

Pisces

The New Moon in Venus indicates that your weakness for men in overalls will soon become public knowledge. The rumours began following a careless remark you made at Christmas when, between bites of Bratwurst, you admitted to a French mustard enthusiast that you 'prefer the flavour of the Colemans'.

Lucky adverb: Meanwhile
Lucky dynasty: Tang

Weekly Forecast for 7th to 13th February

Aries

Take care not to alienate anyone this week – especially your partner, as Jupiter, planet of exasperated eye rolling, loud sighs and tutting, is prevalent in your chart. A man with wild staring eyes and a beach hut in Frinton may try to involve you in a porridge-trafficking ring.

Lucky complaint: Vange
Lucky cornet: Vanilla 99

Taurus

Try as you might, you can't seem to do anything about the stream of vile and depraved phone-calls that have been troubling you so much of late. Luckily, none have so far been traced to your office number. On Friday, you will receive 1,817 Valentine cards. 12% down on last year.

Lucky wine: Soixante-Neuf Du Pape
Lucky length: 21cm

Gemini

Mars in your sign is likely to be the root cause of a 'bit of a domestic' midweek when you come home to find that all of the furniture has been rearranged. One positive aspect of this is that your favourite drinking-saucer reappears.

Lucky treaty: 1821 Franco-Prussian alliance
Lucky nightmare: Wolves under the bed

Cancer

Seventeen pints of Green Chartreuse shandy and a Prawn Bhuna with Eamonn Holmes is never a good way to end a Saturday night, so you might be feeling a little fragile this week. On Wednesday, Pluto opposing your birthsign indicates that you will meet a large man with bushy eyebrows who will pass on a red-hot tip.

Lucky fruit: Avocado
Lucky library: British

Leo

Those awkward clashes with someone near you could turn into a full-blown power struggle on Tuesday when your ruler the Sun is in dispute with Pluto. Try not to worry, as on Thursday you'll be offered a job as lead singer in a Showaddywaddy tribute band.

Lucky philosopher: Thomas Aquinas
Lucky matches: Swan Vestas

Virgo

Uranus has an unusual aspect this week, which might put a bit of a damper on romance. Use the free time wisely. It's high time that you did something more for your body than occasionally slumping over the arm of the chair to burp it. This will also help to ease the pressure sores.

Lucky thread: Whitworth
Lucky lunch: Liver and bacon

Libra

Your ruler turns retrograde on the same day that the Sun moves into Libra, which, as you know, means a recurrence of the bracket-fungus right where your sock elastic goes. On Wednesday the cause is finally tracked down to the traditional hand-painted wooden trousers you wear for the village Goose harvest.

Lucky currency: The Turkish Delight
Lucky decoy: Swanee Whistle

Scorpio

There could be some disruption this week. You may find yourself inundated with visitors, and then discover that a fuse blows or there's a flood. The good thing is that you are at your most inventive, and can rise to the occasion with aplomb as you bought a gross last Christmas.

Lucky blemish: Stigmata
Lucky yearning: Horlicks tablets

Sagittarius

It is a sad fact that today's generation doesn't even seem to know the meaning of longanimity – try not to let it get to you. On Thursday, a rare and wonderful trine between Mars and Neptune means that at long last, it appears that Barcus is willing. On Friday, you will pick up a set of beautifully marked Knopflers at a local boot-sale.

Lucky crisps: Eucalyptus
Lucky foible: Catnip

Capricorn

Your irritable mood looks set to continue until Thursday when you take delivery of a matching pair of garden mood-swings. The good news is that you are at your most inventive, and on Wednesday discover a cure for solar-power.

Lucky motion: Brownian
Lucky collar: Astrakhan

Aquarius

You have always imagined yourself to be popular – someone to whom others look up and admire, a disciplinary force to junior colleagues, and a trusted confident of those in authority. So you would do well to pay heed to rising Saturn if you want to maintain your position – and for goodness sake stop putting your hair up in that ghastly hand-knitted snood.

Lucky marinade: Sherry and soy sauce
Lucky alkali: Potash

Pisces

Neptune is still influencing your mood this week so try not to be quite so sensitive. When the lady in the cake shop asks if you had a 'tiddler' on Thursday, she will be asking for change. Pulling her over the counter by her lapels and providing tangible evidence is probably an over-reaction.

Lucky heron: Roll-mop
Lucky cake: Wayward slice

Weekly Forecast for 14th to 20th February

Aries

On Wednesday, Mars rising indicates that although you may feel on top form, other people might see you as stern, or even dour. An acute angle between Venus and Pluto means a clash with a Chinaman in a knitted frock coat over the bet you were supposed to put on for him on Saturday.

Lucky drupe: The plum
Lucky biscuit: Garibaldi

Taurus

Your ruler Mercury goes trine on Wednesday, which means that you'll make significant progress with that new admin assistant who asked for a slice of your Australian upside-down cake. Play your cards close to your chest until baking day on Thursday when a square Pluto means you should be in with a chance of a larger portion than originally expected.

Lucky trousers: Oxford bags
Lucky warbler: Willow

Gemini

Uranus is particularly unpredictable midweek so allow a little extra time for any kind of group activity that might involve baby-oil and a rotary clothesline. On Wednesday or Saturday you could make a journey that will result in the creation of a semi-professional formation lawnmower display team in West Byfleet.

Lucky gazelle: Thompsons
Lucky surfeit: Lampreys

Cancer

Sneaking a few days off last week on the pretext of having a cold was an inspired move, and allowed you the time to put the finishing touches to your magnum opus *Gordon Brown – The Musical!* On Friday a wonderful aspect between your ruler Mars and dreamy Neptune means that at last that new moisturiser starts to work and you can finally say goodbye forever to flaky earlobes.

Lucky lozenge: Throaties
Lucky quay: Parkestone

Leo

After your well-deserved holiday recreating genuine mediaeval thatching using 'Nonny-Nonny Hay', you will be anxious to get back to your usual 2 hours solid work a day. The rocky outcrop you passed on your way home is nothing to worry about and was probably due to the free-range eggs.

Lucky Oxide: Aluminium
Lucky line: District

Virgo

Refreshed from your recent triumph (a 1300cc brown Toledo) you can't wait for the next challenge, and it seems that with Neptune in your birthsign on Tuesday you won't have long to wait – you can't reach your shoelaces again. Mercury transiting your sign on Thursday indicates that despite what you've been told, Matabeleland is not a theme park after all.

Lucky fencing: Waney lap
Lucky vegetable: The eddo

Libra

Even if you have had some difficult times recently, in work and at home, this week is looking positive – apart from Wednesday when mischievous Pluto indicates that you will be unmasked as the head of an international angelica smuggling ring and have to flee the country. Remember your ointment on Tuesday when Venus moves into your sign aggravating your dry gulch.

Lucky movement: Pincer
Lucky snack: Malt-loaf

Scorpio

Many people with the Sun strong in their charts may find their career can advance now. This however does not apply to you as at the end of the week you discover that you are being reared for your valuable pelt. You may need to leave a ladder in the bathroom on Tuesday as you discover that in your case cleanliness is next to weightlessness.

Lucky heath: Haywards
Lucky trousers: Translucent

Sagittarius

On Tuesday you will have cause to remember the Sicilian waiter that you upset last week while demonstrating a 'Basildon blow-torch' in his restaurant, and as a result you find out that a Vendetta is not in fact an ice-cream dessert. Mars is transiting Uranus on Wednesday, so be prepared for a chilly reception.

Lucky constant: Planck's
Lucky Paige: Elaine

Capricorn

On Tuesday, a trine Pluto means that the number 17 bus will be hijacked by Welsh fundamentalists again. Your intimate knowledge of *Men of Harlech* serves you well. Toward the end of the week, you will receive good news about your application to play banjo for the Royal Philharmonic.

Lucky attire: Red feathers and a 'hooly-hooly' skirt
Lucky stamp: Petulent

Aquarius

Saturn rising indicates that on Wednesday afternoon, you will meet a stranger in a green jacket. Under no circumstances should you buy tinned carrots from this person. A careless comment in the betting shop on Friday may mean that you accidentally become Editor-in-Chief of the *Jewish Herald*.

Lucky regiment: The 17th/21st Lancers
Lucky hairstyle: The Beehive

Pisces

An unusual start to the day on Tuesday, when you are woken by Lord Soames dressed only in a short silk dressing-gown doing star-jumps in front of your picture window. The New Moon rising will mean a domestic mishap on Thursday, when you find that the box of moist toilet tissue on the cistern turns out to be bleach-soaked bathroom wipes. This could well provide unexpected highlights.

Lucky tipple: Lager-tops
Lucky phrase: For goodness sake, not in the sink!

Weekly Forecast for 21st to 27th February

Aries

On Monday a benevolent Jupiter will bring wellbeing and good fortune in matters relating to your fine collection of Victorian ephemera. On Thursday you will finally break your cardigan habit when your doctor prescribes elbow patches. On Thursday you should try to avoid women with split-ends who shout too much.

Lucky nebula: The horse-head
Lucky break: 147

Taurus

Recently, you've not only been taking care of your own affairs, but helping to plan other people's. This could well turn into a lucrative business. On Friday, Venus and Mercury both aspect Pluto which will lead to thc discovery that you don't enjoy Lacrosse very much.

Lucky tipple: Pink gin
Lucky Stone: Sharon

Gemini

'Gemini' was once a code word given to Patrick McGoohan in *The Prisoner*. Using it brought him not safe passage but a comprehensive beating. You can expect much the same result if you continue to promote your new idea of 'pay per view' ladies foundation garments in the office. Wednesday's Venus in your fifth house means that your ability to play the 'spoons' solo from *Swan Lake* may lead to a recording contract.

Lucky state: Supine
Lucky chord: C Major 7th with a diminished 9th

Cancer

Tuesday could bring an unusual offer from a man in a raincoat with sweaty palms. Before you commit to anything, check the small print. Your ability, in an emergency, to make a noise like a dolphin serves you well. Wednesday's lovely trine between Saturn and Neptune indicates a mild but persistent ear infection.

Lucky jump: The triple-lutz
Lucky sandwich: Brie and avocado

Leo

Jesus may well want you for a sunbeam, but a square Saturn this week indicates that big Rosa from the Italian Restaurant has altogether more down-to-earth plans for you. On Wednesday, troublesome Pluto will mean that you are caught on camera plucking your ear-hair and stuffing it into your wallet during the National Anthem.

Lucky dwarf: Sneezy
Lucky accent: Hungarian

Virgo

Wednesday's Jupiter rising may go some way toward helping you unravel the meaning of your recurring dream about Peter Andre gargling with white mice. The New Moon in Neptune will mean that on Wednesday you will be offered the role of Belstaff in a modern-dress version of the *Merry Wives of Windsor*. It's a similar role to Falstaff, but requires you to ride a Norton Commando.

Lucky Humber: Super Snipe
Lucky vitamin: G-Major

Libra

Like many with Venus strong in their charts, you have a bit of a reputation to uphold. A square Pluto on Monday means that something you have dreamed of for some time finally comes true when you are invited to show your peonies to the Womens' Institute. It might be best to get the request in writing to avoid any misunderstandings.

Lucky fireplace: Adam
Lucky card: Get well soon

Scorpio

Toward the end of the week, mysterious Neptune transits your birthsign, and as usual, this indicates another run-in with the authorities over catching cats with a rod and line out of your bedroom window. On Friday, square Saturn means that you will no longer go unrecognised, as you will have a root vegetable strain named after you.

Lucky borough: Enfield
Lucky tipple: Sanatogen Sling

Sagittarius

Neptune forms an interesting angle to Mars in your birthsign later this week, which means that your contact lens fitting will take a turn for the worse, when one of them accidentally slips behind your eye, gets into your bloodstream, and leaves you with a permanent inability to finish DIY projects.

Lucky bearing: North by Northwest
Lucky weakness: Marzipan

Capricorn

As you are reminded all too frequently, Uranus often gets you into trouble. This week is no exception. On Thursday, your regular game of British Bulldog with the Benedictine Sisters is interrupted by a ferocious gust of wind (for which you are blamed) that carries off the local Royal Mail records, leaving many people in the area without Postcodes. On Friday you receive an unsolicited seed-cake.

Lucky bull: Papal
Lucky sauce: Marie-rose

Aquarius

An eventful week is indicated by a trine Saturn on Wednesday. Even though your nightmares about Sir Alan Sugar continue, things will improve toward the end of the week when you come under the benign influence of Venus. On Friday, you will discover that the combination of Absinthe and mushy peas makes you glow in the dark.

Lucky pet: Terrapin
Lucky fritter: Pineapple

Pisces

A square Neptune this week will spell trouble for you on Tuesday when a busy-body points out that your new commercial venture in Conqueror-bond toilet tissue has one fatal flaw. On Thursday a new acquaintance may be struck down in Broadstairs by a falling domestic appliance. This may be a blessing in disguise.

Lucky pastry: Puff
Lucky condiment: Mint sauce

Weekly Forecast for 28th February to 6th March

Aries

A slow start to the week, but on Friday the Sun challenges your ruler leaving you with a profound uncertainty about the fashion-sense of the Deputy Prime Minister. Shouting through his letterbox will not make him see sense either. On Thursday, Chiron transits your birthsign, which is just typical. Avoid collards.

Lucky marmalade: Quince
Lucky unit of measurement: A perch

Taurus

Even though you are well known as one who barely eats enough to keep body and soul together, Tuesday's waning moon indicates a lunchtime mishap – you may well end up with a toasted panini – particularly if you stand too close to the grill. On Thursday, you get a call from Kew Gardens to ask if they can name a new type of mildew after you.

Lucky stork: Maribou
Lucky sunbeam: Rapier

Gemini

Once again Mars in your birthsign may lead to an overwhelming urge to take up pigeon racing. Resist it. It can only mean heartache – just as it did the last three times. A large man with a midlands accent, a packet of Fruit Polo, and a tartan shopping trolley may once again try to take advantage of your kindly nature on Wednesday lunchtime. This time obey your instincts and have him neutered.

Lucky tempo: Largo
Lucky pickle: Lime

Cancer

As you know, you have Saturn trining Neptune in your natal chart this week. This indicates seriousness, sobriety and responsibility to obligations, and someone who is not generally prone to having their hair dyed blue and faxing photocopies of their personal regions to prominent members of the European Parliament – so give it a miss this week. On Thursday you may be melted-down for scrap.

Lucky pie: Chicken & mushroom
Lucky glue: Copydex

Leo

This month continues well for you on almost every front. Romantically, it could hardly be better. Between now and Thursday, when Mercury goes direct, you will discover why Dentists voluntarily wear 'star-trek-style' jackets that button down one side, and why there is always a red-rubber workman's glove on every roundabout. Jupiter's entry into Virgo indicates Kedgeree by the weekend.

Lucky bat: Pipistrelle
Lucky trousers: Plus fours

Virgo

On Tuesday, Mars transiting your ruler, the Moon, indicates that your water on the knee will take a turn for the worse when the Doctor tells you that you're three months stagnant. Later in the week, Saturn enters your fifth house leading inevitably to another jellied-eel binge.

Lucky pancake: Buckwheat
Lucky song: *Eskimo Nell*

Libra

An unusual square between Jupiter and Pluto means that on Wednesday, you will discover that your best trousers are haunted after they begin making a high-pitched whooping noise whenever you meet someone with red hair. Your deltoids are likely to remain troublesome until Mars goes retrograde at the weekend.

Lucky manifestation: Golden shower
Lucky draw: Bolton Wanderers

Scorpio

Communication is difficult this week, as you spend Tuesday trapped in a lift with Boris Johnson. By chance, the arrival of Neptune in Leo will give you the uncanny ability to follow his train of thought. Saturn has just entered your birthsign and has brought with it a deep concern about lawnmower maintenance that will give you restless nights until the middle of next week.

Lucky lock: Teddington
Lucky canal: Grand Union

Sagittarius

Well, Mercury is retrograde this week, which means that poor weather will cause the cancellation of all performances at the Globe. You and several colleagues may be called upon to act as a theatrical 'pools panel'. Just put Romeo and Juliet down as a score draw, and Richard III as an away win for the Tudors. On Friday, a fried-egg sandwich proves troublesome and may have unforeseen consequences later in the day.

Luck solvent: Acetone
Lucky orchestra: Royal Philharmonic

Capricorn

Mercury, planet of communication, is retrograde in Leo this week, which means that nagging doubts, self-recriminations, bitterness, and past embarrassments will give you troubled nights until Thursday, when you are beaten up by Sir Clive Sinclair in the Co-Op after you inadvertently whisper to him 'Together Vladimir, we shall dance until dawn'.

Lucky wine: Watermelon
Lucky dumpling: Parsley

Aquarius

A hilarious incident in Luton with a flock of Canada geese and a tumble-dryer changes your fortunes on Wednesday. A retrograde Pluto means that Waterfowl will continue to play a part in your life later in the week when your recipe for 'duck mystique' wins you the undying admiration of Delia Smith.

Lucky highlander: Gordon
Lucky currency: The Danish Pastry

Pisces

The Full Moon in your sign at the weekend promises that despite early predictions, your second publication, *Knife-fighting with Sister Wendy* will make the best-seller lists next month. Mercury, planet of communications is trine with Mars on Wednesday, so throwing hedge-clippings out of your bathroom window has to stop.

Lucky inflatable: Peter Mandelsson
Lucky flatfish: Dover sole

Weekly Forecast for 7th to 13th March

Aries

Mercury comes in many guises, but rarely in the form of the silver robots from the Smash commercials, so you can put that one down to the after-effects of the homemade parsnip wine on Sunday. The ongoing creative quintile between Uranus and Pluto indicates that your chequered past actually turns out to be a rather fashionable Burberry. A huge man from Hammerfest may approach you with an unusual suggestion on Thursday. Do nothing that involves the word 'dunnage' if you know what's good for you.

Lucky Fife: Robertson
Lucky harness: Martingale

Taurus

Your ruler, Venus, is now travelling hand in hand with Pluto, which means another late-night confrontation in your back garden with the camouflage section of the band of the Coldstream Guards. The traps didn't work last time, and they won't this time either. On a brighter note, the trade-deficit falls.

Lucky lectures: Reith
Lucky owl: Tawny

Gemini

If you are considering moving to a larger house, the Sun forming a trine to Saturn on Monday will lead to a disagreement about whether or not you should take carpet tiles with you when you move. Mars rising on Friday means you will rediscover the delight you once took in Cream of Tartar.

Lucky game: Ptarmigan
Lucky hedge: Privet

Cancer

Mars is square your ruler the Moon so this week may get just a little too hot for comfort, particularly for anyone working with the media, or just those with a dark secret involving wallpaper paste. On Tuesday, you will have a chance encounter with someone who has 'an unusual way' with courgettes that will both surprise, and delight you.

Lucky element: Bismuth
Lucky mast: Mizzen

Leo

Life has been unexpectedly quiet of late, but all that is about to change on Wednesday when a square Saturn leads you to discover a new culinary delight at a favourite lunchtime haunt. Your elation is likely to be short lived, as fickle Pluto indicates that what you read as 'Corned Bee' may well be a printing error.

Lucky livestock: The Alpaca
Lucky Motion: Andrew

Virgo

Venus, the planet of love, moves into your own sign on Tuesday, however, the path of true love seldom runs smooth, and in your case, can have unpredictable twists and turns – even before you mention articulated shoe-trees and creosote. On Friday you will be mistaken for Stewart Granger at the fish counter in Sainsbury's and forced to autograph a series of tartan shopping trollies.

Lucky utensil: The spatula
Lucky costume: Doublet and hose

Libra

A difficult week that starts badly, then really goes downhill from there. Pluto's malign influence is particularly felt on Wednesday in the staff canteen when a gooseberry surprise is particularly potent and makes you spill oxtail soup in your handbag. You will be followed home by a relentless Schnauzer.

Lucky blemish: Hives
Lucky shears: Crimping

Scorpio

Mysterious Neptune is rising in your fourth house on Wednesday, so when an attractive lady with an Eastern European accent and a bag of lemon bon-bons approaches you in Waitrose with the words 'While your shoes are stretching, I will dance the Polka with you', give her a tin of black shoe-polish, she will know what to do. On Friday you will be sprayed with perfume in a random attack by a gang of Yardleys.

Lucky bag: Gladstone
Lucky sword: Epee

Sagittarius

Uranus dominates the early part of the week, and things are still unsettled for you at the weekend when foreign travel is indicated by the passage of Mercury planet of communication through your birthsign. On Thursday lunchtime you will meet a man who will teach you how to say 'it's coming out like oxtail soup' in seven languages. This could prove more useful than you might imagine.

Lucky tie: Newcastle and Everton
Lucky Ball: Zoe

Capricorn

Your ruler, the Sun, is now in Sagittarius. It's much more comfortable here than it was in Scorpio, so don't take any more if its old nonsense. Some startling new information about the lifecycle of Wombats could come to you this week but you won't recognise it for what it is so an opportunity will be missed. Still... Never mind eh?

Lucky ironing board: Minkey Starlite
Lucky lemur: Ringtailed

Aquarius

Somewhere in your circle of friends and family is someone with whom you've fallen out. This might be a longstanding coolness, or a just a misunderstanding about the incident with the Verger, a chocolate digestive, and the Durham Light Infantry. On Tuesday, you will be attacked in the tabloids over your extreme views on the piano-accordion.

Lucky ballroom: Hammersmith Palais
Lucky duck: Muscovy

Pisces

You're no stranger to hard work. You are used to making plans and keeping doggedly on with them in circumstances that would make weaker spirits quail. Pluto rising indicates that this has at last been recognised and your name has been put forward for a big job in Rome that involves public speaking, tarmac, and tucking in your shirt with a wooden spoon.

Lucky dog: Borzoi
Lucky lake: Veronica

Weekly Forecast for 14th to 20th March

Aries

Neptune transits your birthsign at the beginning of the week bringing with it true clarity of thought. On Tuesday when you are putting out the bin, you suddenly realize that there are similarities between this action, and your social life – only taken out once a week, in the dark, by a dustman. It may be time to freshen up that wardrobe.

Lucky vegetable: Celeriac
Lucky equation: Simultaneous

Taurus

As Saturn moves into your fourth house on Tuesday you receive a postcard from a friend on holiday in California and are so inspired by the picture that you decide there and then to resurrect your flagging musical career with a bagpipe instrumental of the Tony Bennett classic – *Don't mess with my toot-toot*. Don't be too optimistic though as a square Neptune means that the video will be rubbish and the single will only get to number 22.

Lucky cartoon: Bleep & Booster
Lucky font: Garamond

Gemini

Mysterious Neptune is square to your birthsign this week, which may help to explain why, since Christmas, you have lost the ability to walk silently in corduroy trousers. Although you will lose weight over the next few weeks, it will be only from your left side. On Tuesday afternoon Pluto rising means that the itching will get worse, so pick up a Dutch hoe on the way home.

Lucky Realm: Woman's
Lucky sneer: Supercilious

Cancer

Your enthusiasm for Karaoke remains undiminished but this may be your undoing as a trine between Mars and Venus indicates that on Thursday you will be secretly filmed on stage putting everything you've go into *The Girl From Ipanema*. This may lead to a sudden career change. The New Moon in Aquarius indicates that Sardines on toast will make a welcome return as a mid-morning snack.

Lucky profession: AA Patrolman
Lucky convulsion: Febrile

Leo

A disturbing aspect between the Moon and Mercury will lead to confusion early in the week as you will make a moonlight discovery with your shin that all your furniture has been rearranged. You will have no idea how, or why this happened. Mars is square to your birthsign on Wednesday, which as usual, means you will suffer catastrophic and sudden clothing failure. Try not to over-react.

Lucky iris: Bearded
Lucky grape: Zinfandel

Virgo

A square aspect between your ruler, the Sun, and Jupiter will mean that plans for your walking holiday in Austria are going awry. Not only will you discover that Sturmey Archer don't make mountain boots with low-ratio gears, but that damage sustained to your Melodica in last year's ascent of K2 was the result of Chinchilla action, and so not covered by insurance.

Lucky sausage: Bratwurst
Lucky fish: The Wahoo

Libra

Jupiter rising means that career issues are highlighted this week. With a lot of change going on around you it might be time to reconsider whether turning down the top job last time around wasn't a mistake. An unusual trine between Mars and Neptune on Thursday could indicate that you are being watched by aliens, which should please your employer.

Lucky vegetable: Rocket
Lucky medication: Sloanes linament

Scorpio

With Mars in opposition to Saturn this week, there will be an incident on Wednesday in the High Street involving a running machine, a pan of clarified butter, and a seafood delivery van. You will receive a severe ticking off from the traffic Police, and two Michelin Stars for the creation of what will come to be known as Lobster Esplanade

Lucky tense: Future imperfect
Lucky walls: Jericho

Sagittarius

On Wednesday afternoon, a trine Mars means that you will have a misunderstanding with a second hand television dealer when you ask if he'd be interested in looking at an old black & white Bush. On Thursday, a man with piercing breath and a Renault 4 may enter your life. Under no circumstances should you agree to share your gravy browning with him.

Lucky verb: Dangling participle
Lucky vault: Diamadov turn

Capricorn

Be patient if a romantic relationship is not developing the way you'd imagined. The equipment is unwieldy, and sometimes difficult to operate for a novice – particularly those from the stricter religious orders. At the end of the week, Saturn moves into the empty house next door but one. Pop around for a cup of tea and a bit of a chinwag.

Lucky pancake: Crispy duck
Lucky look: Tousled

Aquarius

After a quiet weekend hiding in the coal bunker things will start to get altogether more lively on Tuesday when you are swept along by a mob of anarchic pensioners on their way to Downing Street. By the time you arrive, you are the only one who remembers what their demands are. As their spokesman, you are immediately arrested for inciting a mob. On Friday, the cold floor of the cell will give you trouble with your 'Rockfords'.

Lucky suspension: Transverse leaf-spring
Lucky fabric: Horizontal stripes

Pisces

This afternoon there will be some confusion in Waitrose when you ask an assistant if they have any tinned pears in stock. They will only have syrup, so you may have to settle for cling peaches in a rich onion gravy as usual. A square between Mercury and Pluto on Wednesday means that the usual Friday night 'bar-maggot-race' at the local pub may be cancelled due to an influx of the 'wrong sort'.

Lucky polymer: Long-chain
Lucky snack: Bread-pudding

Weekly Forecast for 21st to 27th March

Aries

The Full Moon in Mercury means a crisis of confidence on Wednesday afternoon when you suddenly find that you've lost the ability to tell the difference between Stork and Butter. Swift action by a passing Rosecrucian will soon rectify the problem and you'll be back to normal by the weekend.

Lucky envelope: DL
Lucky cough: Throaty

Taurus

The waning Moon in Mars means that on Wednesday you will receive a surprise call from the new Wales Manager, who, having checked your credentials, offers you a game at Fly-Half on Saturday. Unfortunately, you will have other plans, as a dodgy prawn Bhuna on Thursday evening will see you taking very small, careful steps to recovery.

Lucky composer: Prokofiev
Lucky river: Limpopo

Gemini

The New Moon in Pisces indicates a busy time. You will start the week in a whirl of activity, but on the whole it is positive and productive until Thursday when a quick lunchtime drink turns into a three day Cinzano-fuelled binge and you wake up in Lowestoft with a tattoo of Bernard Matthews on your left thigh, and an uncontrollable addiction to whelks.

Lucky publication: *The Guinness Book of Revelations*
Lucky snack: Vesta Chow Mein

CANCER

A restless start to the week is indicated, particularly following the weekend Saturn-Uranus. On Wednesday, you will finally discover the reason for your reluctance to abandon the bachelor lifestyle when you are diagnosed as a curable romantic. You'll be particularly clumsy on Thursday, so the 'Jack Douglas' 731-piece tea service might prove a good investment after all.

Lucky disguise: Reg Varney
Lucky shoe protectors: Blakeys

LEO

A part-time job, or hobby may reach its natural conclusion this week. A fascinating trine between mysterious Neptune and Uranus, planet of surprises will mean that on Tuesday you will get into a violent quarrel about pickled cabbage with a truculent bottle-blonde. Swift action by a passing Quantity Surveyor will save the day and the bruises will start to fade by the weekend.

Lucky sensation: Tingling
Lucky marmalade: Grapefruit

VIRGO

Nobody knows the trouble you've seen. Except for Sharon at number 16, that is. You know – the lady with the poodle and the binoculars? However, Saturn rising on Wednesday will lead to an immediate improvement in your fortunes, as you will receive good news about the screaming noise from your airing cupboard, which is not a faulty thermostat as you feared, but a bogeyman, which even terrifies the wolves under your bed.

Lucky bird: Corn-Grunting
Lucky hinge: Rising-butt

Libra

Thursday's Pluto square to your birthsign means that while digging the foundations for a new Trebuchet, you will discover an ancient stone tablet on which mystic runes are carved. Oddly enough, when you finally find someone to translate them from ancient Ogham, it will turn out to be a copy of the third edition of the Des O'Connor songbook.

Lucky cartoon: *Noggin the Nog*
Lucky relative: Great Aunt

Scorpio

With Mercury high in the section of your chart that governs your career you should buy that safari-suit you've been admiring for weeks – A smart outfit will come in handy when you need to start looking for another part-time job. You may experience a setback towards the end of the week, when you find out that despite a positive test, it is squirrels.

Lucky hash: Corned-beef
Lucky nut: Three-eighths Whitworth

Sagittarius

Those awkward clashes with someone near you could turn into a full-blown power struggle on Tuesday when your ruler the Sun is in dispute with Pluto over your plans for the launch of *Ear-Hair Monthly* magazine. Try not to worry, as on Thursday you'll be offered a job as lead singer in a Steps tribute band, which will tide you over.

Lucky banjo: 5-string
Lucky order: The Carmelites

Capricorn

This week, Venus, planet of love is in the ascendant, and this combined with mysterious Neptune, will make you utterly irresistible to the opposite gender. On Wednesday, a shopping trip will nearly end in tragedy when two lady shop assistants catch a fleeting glimpse of your profile and burst into flames like the map at the beginning of *Bonanza*.

Lucky precaution: CO2 Fire extinguisher
Lucky number: 999

Aquarius

On Tuesday, unpredictable Aquarius is about to conjoin with Neptune in the area of your chart governing infantry tactics. On Thursday you will receive confirmation that despite a positive test, it is unlikely that lizards could have caused the condition. The good news is that when it has been carefully washed and dried using a lint-free cloth, the ocarina should once again function normally.

Lucky dance: The Hustle
Lucky chutney: Plum

Pisces

At the beginning of the week Saturn and Pluto are once more in exact opposition and I'm afraid that even though you usually have all the answers, you may struggle to explain your actions in the National Portrait Gallery on Tuesday lunchtime. Try to remain calm and remember that despite all arguments you hear to the contrary, if it's over four inches long, it's technically a Sardine.

Lucky soup: Gazpacho
Lucky veneer: Figured walnut

Weekly Forecast for 28th March to 3rd April

Aries

As Thursday's Full Moon approaches, a man with dense eyebrows may bring you good news about your quest for another part-time job, although disappointingly, the paraffin round which you applied for last Wednesday will have gone to a younger man. On Friday, you will develop an irrational hatred of sideboards.

Lucky bandage: Triangular
Lucky process: Osmosis

Taurus

A square Saturn means a bit of a mixed bag fortune-wise. A rash purchase of a packet of iced-gems means that you will need to visit the dentist to have the hole in the roof of your mouth repaired. You may be jostled in the lift by David Cameron who will try to sell you a set of used cummerbunds.

Lucky wave: Sine
Lucky ointment: Bonjella

Gemini

This week there will be a wild-west air about the home. You will be forced to confront your partner after a series of restless nights during which you are woken repeatedly by loud 'mooing' noises from the bathroom. On investigating, you will find a herd of Longhorns with no rational explanation for their presence other than, 'I found them on the way home'.

Lucky vessel: Spitoon
Lucky pine: Lonesome

Cancer

This week you are offered the lead role in a Bond-style remake of a Greek tragedy, *Oedipussy* in which you are involved in a love-triangle with Isosceles and Pythagoras, but a protracted argument about camera angles halts the production. Try to remain stoic.

Lucky rhythm: Samba
Lucky lubricant: Lard

Leo

There may be a misunderstanding about a gift this week. So, even if someone offers you a pearl necklace, make sure you know what is expected of you before accepting. Toward the weekend, troublesome Mars means that you won't get that hole in your best socks darned again. Those with Saturn rising may dream about bats with beards.

Lucky colour: Puce
Lucky snack: Black grape and brie ciabatta

Virgo

The week starts with the New Moon in an area of your chart that inclines you toward night-cramps. There may be a misunder-standing in the butchers on Thursday when you are mistaken for Carol Vordermann. Try to keep your dignity intact.

Lucky garnish: Parsley
Lucky prank: Whoopie cushion

Libra

Sunday's New Moon means that during community whistling, you'll probably forget the opening two bars of Schubert's *Marche Militaire* again. But it will also bring you an unexpected piece of good fortune in that no-one will be listening anyway. You may be in trouble over forgetting to oil those gate hinges.

Lucky shoes: Black brogues
Lucky ligament: Anterior Cruciate

Scorpio

A frustrating week starts off with a fruitless quest to track down Cyrillic Alphabetti-Spaghetti on Wednesday. The forecast for the weekend isn't much better with variable 3 or 4 at first in Northwest Rockall, occasional drizzle, moderate to good. On Friday, you will be both surprised and delighted with the range and flexibility of frozen Hake loins in the Co-op.

Lucky font: Times New Roman
Lucky affliction: Shepherd's Bush

Sagittarius

It was only a matter of time before someone found out about your collection of celebrity nose-hair. The best you can now hope for is that they don't find the toenail clippings in your vanity case. A tall stranger may break in and do all your ironing.

Lucky potato: Maris Piper
Lucky haircut: Mohican

Capricorn

This week sees you on a bit of a roller coaster. Alton towers have increased their payload by 50%. Mercury is moving into dynamic Aries, which means your mower is now overdue for a service. Thursday's Full Moon indicates that you could suffer from some gossip about you and yours. Just ignore it, you still have the negatives – and the egg-whisk.

Lucky literary device: Metaphor
Lucky accent: Welsh

Aquarius

Uranus has just moved into Gemini, so on Tuesday you can expect to become the new 'face' of Revlon Cosmetics. Be extra careful with root vegetables especially after that incident last week – you don't want to have to explain that 'fall in the greenhouse' to your doctor again.

Lucky shade: Violet
Lucky whelk stall: Tubby Isaacs

Pisces

An interesting aspect between Venus and Pluto indicates that you should get some exercise, but not the way you've been doing, which is not only ethically wrong but technically illegal. Mercury is in Taurus, so don't take what the butcher said seriously-he doesn't even know the Club secretary.

Lucky toothpaste: Aquafresh
Lucky craft: Knitting

Weekly Forecast for 28th March to 3rd April

Aries

With Mercury high in the section of your chart that governs your career you should buy that tank top you've been admiring for weeks – A smart outfit will come in handy when you need to start looking for another job. You may experience a setback towards the end of the week, when you find out that despite a positive test, it is only water retention after all.

Lucky pudding: Blancmange
Lucky accessories: Snorkel & flippers

Taurus

Venus is retrograde in Mercury this week, which means that your boss could well deliver a bombshell on Wednesday. Fortunately a trine Saturn means that the fuse is likely to be faulty and it won't go off. On Friday, a woman with overly stout body hair will show you her gazebo.

Lucky enzyme: Insulin
Lucky panto: Mother Goose

Gemini

At long last things for Gemini are looking up. On Tuesday, a lovely aspect between the Sun in Virgo, related to Gemini by the common rulership of Mercury in the area of your chart affecting foot-care means that you will finally sort out that bunion once and for all. Despite advice to the contrary, loon-pants are not staging a comeback.

Lucky utensil: Ronco AutoChop
Lucky facial expression: Bemused

Cancer

Your ruler, The Moon, is now travelling hand in hand with Uranus, planet of surprises – so you could be accident-prone. On Thursday, a man with welder's gauntlets and a box of herrings may shout at you from the bushes. Try not to let him affect your backswing.

Lucky interjection: 'Huh!'
Lucky Womble: Orinoco

Leo

Mercury goes retrograde on Wednesday, which for you signifies the beginning of a new phase in your life. Unfortunately, it's just like the old phase, so no-one will notice. Saturn rising indicates that your new teeth will be ready on Friday, just in time for 'celery night' at Stringfellows.

Lucky olive: Kalamata
Lucky religion: The Rosicrucians

Virgo

Just for once, try to voice your feelings this week. Let people know what you think for a change. On Wednesday, a gooseberry surprise will turn out to be rather dull and predictable. On Friday, a white-haired man with a sparse beard will attempt to stow-away in your hand luggage. Try to ignore his plaintive cries for cheese.

Lucky lunch: Tripe & Onions
Lucky currency: The Vietnamese Dong

Libra

This week fickle Pluto thwarts your attempt to become the world's strongest woman. On Thursday the Duke of Edinburgh passes on some good news about your search for reduced-calorie chewing-tobacco. At the weekend you should keep a tight rein on your urges as Ricky Martin is in town, and he hasn't forgotten Wantage even if you have.

Lucky composer: Erik Satie
Lucky pickle: Red cabbage

SCORPIO

Rising Mercury suggests you need more change in your life right now. Try checking down the back of the sofa. On Tuesday a square Jupiter indicates that there may be trouble ahead. But while there's moonlight and music and love and romance, just wear your blue nylon pants.

Lucky Clipper: The Cutty Sark
Lucky fielding position: Fine-leg

SAGITTARIUS

On Thursday, a rare and wonderful trine between Mars and Neptune is perfectly placed to enable you to sleep right through the alarm and turn up for work looking like one of those men you see in the park talking to bottles of cider. Beware of those ground-hugging mists you get in shampoo commercials.

Lucky feature: Hazard lights
Lucky taxi driver: Brian

CAPRICORN

The Full Moon in Mercury means a crisis of confidence on Wednesday afternoon when you can't decide between the Belgian Bun and the Eccles Cake. On Friday the back of your legs will be severely slapped by a forthright lady colleague with firm views on summer bedding. Uranus rising indicates that your contract-flooring business will continue to go from strength to strength.

Lucky ailment: Travelling Wilburys
Lucky instrument: Endoscope

Aquarius

Money is a little tight this week, and with the Full Moon in challenging Saturn on Wednesday, it looks like you'll have to wait another month for those tyres for your Sprite Wayfarer caravan. This will come as a bitter blow to your partner who was particularly looking forward to Prestatyn next week.

Lucky instrument: Timpani
Lucky cartoon: Bleep & Booster

Pisces

An interesting aspect between Saturn and Pluto this week could mean that you are contemplating an image change. Having always been a slave to fashion, it will come as something of a surprise to your colleagues to see you giving up the Versace leather catsuit. On Wednesday, a chubby man with abundant dandruff will give you a good tip for the 4.10 at Kempton Park.

Lucky mammal: The Agouti
Lucky allsort: Liquorice

Weekly Forecast for 11th to 17th April

Aries

Volatile Neptune enters your third house on Tuesday when you leave the bathroom window open. This leads to a swarm of masonry bees taking up residence in the equipment closet and refusing to leave until you give them the correct handshake.

Lucky cardboard: Corrugated
Lucky fabric: Senegal tweed

Taurus

An interesting week during which new opportunities present themselves. A square Pluto indicates you'll make significant progress toward converting coal-tits to smokeless fuel. On Wednesday, Mars entering your sign means that you make the surprising discovery that you are allergic to vowels.

Lucky tribe: Commanche
Lucky diagram: Venn

Gemini

This week's square Saturn means that it's payday again. Were it not for the fact that your Mother still makes you write a 'thank you' letter to the company chairman every month, you'd have very little to concern you. On Friday, a trine Mercury leads you to discover a unique way to separate Siamese cats.

Lucky handshake: DTs
Lucky craving: Apple doughnut

Cancer

There may be a misunderstanding about your forthcoming lead role in the costume drama 'Marge of the Light Brigade'. You thought you were getting paid, the producers thought you were a stalker. The Full Moon is in your birthsign on Monday, so there!

Lucky range: Aga
Lucky statue: Rodin's 'The Thinker'

Leo

Once again, troublesome Mars means that you are unlikely to track down the proof of the pudding this week, and must wait until Tuesdays waning Moon to discover the current position.

On Friday you will be taken to task about your views on the use of skiffle as a muscle relaxant.

Lucky bandage: Crepe
Lucky game: Shinty

Virgo

Saturn is retrograde this week, so there may well be a bit of trouble over the storage of the life-sized ice-sculpture of Hugh Scully in the canteen freezer. Although your studies are progressing well, try not to let the chloroforming get out of hand as people are beginning to talk.

Lucky publication: *The Guinness Book of Matches*
Lucky memory: Eight trilobites

Libra

You can learn a lot from observing the events at the weekend, when your ruler Mercury turned retrograde. The straddle-lift was a rather rash purchase. On Wednesday a warm romantic encounter with a close-order drill enthusiast may be spoiled by wearing a regimental tie to which you are not entitled.

Lucky gesture: Mexican wave
Lucky lance: Percival

Scorpio

On Tuesday a trine Mars means that you'll step boldly forth into unfamiliar and unplanned territory. Other people's mistakes will mean that everything you've done recently has to be reorganised. On a brighter note, your recipe for whelk soufflé will be a huge hit at the Lord Mayor's banquet.

Lucky shrub: Cotoneaster
Lucky calibre: 7.62mm

Sagittarius

With your ruler, the waning moon, in your fifth house, there can be only one course of action open to you. Even though your stars granted you dense body hair, you never dreamed that your own body-topiary salon would be anything other than a wild fantasy. Make that dream come true and you might be amazed where it might lead.

Lucky pudding: Blancmange
Lucky rub: Algipan

Capricorn

Uranus has been square for some time, so it should come as no surprise to you that the combination of an over-rich diet, and sedentary lifestyle is not helping matters. On Tuesday you will get into a heated argument about the best way to transport a flock of show-goats on the tube. As usual, you know best.

Lucky dance: The sailor's standpipe
Lucky precaution: Spare elbows

Aquarius

An impromptu family Eisteddfod on Wednesday may provide an opportunity to get back at your neighbour after he accidentally scratched your second best credenza when he borrowed it last weekend. Thursday brings the disturbing news that your mobile lady chiropodist has her mind on higher things.

Lucky waltz: Disney and Whitman
Lucky snack: Malt-loaf

Pisces

Venus and Neptune are offering you a fairy-tale experience this week. You will find the role of Billy-Goat Gruff a strangely familiar one. Monday's moon in your birthsign indicates a watershed. The roof is leaking in the garage again. Cut-price plastic sheeting won't do the job this time either.

Lucky rhythm: The rumba
Lucky Fish: Michael

Weekly Forecast for 18th to 24th April

Aries

A well-rounded look to the week as Venus and Mercury both aspect Pluto. This means that you really need to take stock. An oxo cube just after lunch should do the trick, but chew it slowly and remember to remove the silver paper.

Lucky footwear: Waders
Lucky cake: Caraway seed

Taurus

Your impending birthday may leave you feeling somewhat confused and are unsure of how you see yourself and your role. With Mars rising, this is perfectly normal. And as the old saying goes, if the shoe fits – you're extremely lucky because mine pinch like the devil.

Lucky furniture: Ottoman
Lucky card: Seven of Diamonds

Gemini

Pluto rising indicates conflict at the start of the week when you return from a *Treasure Island* themed fancy-dress party to discover that you have left the clipper full of corsairs after you used it on your buccaneers. Careful with the hook.

Lucky lamp: Davey
Lucky cable: Vince

Cancer

As a true Cancerian, thoughts of the sea are never far from your mind. The New Moon in Pisces indicates that your attempts to corner the lucrative wholesale shellfish market south of the river will receive a boost this Wednesday when most of the major distributers will agree that they'd like you to handle their winkles.

Lucky hinge: Rising butt
Lucky Starr: Ringo

Leo

An interesting aspect between Jupiter and Venus will leave you in a bit of a spin on Wednesday, when you discover that a distant Welsh relation has left you a controlling share in Swansea's only adult publication *Men of Harlech Only*. On Friday, you discover to your distress, that a niece has entered you for the Ask-Aspel Challenge Trophy.

Lucky Clampett: Elly-May
Lucky tribe: The Arapahoe

Virgo

You may experience a setback towards the end of this week, when you find out that despite a positive test, the rate of head-line inflation has nothing whatsoever to do with the noises in your left ear. On the plus side, a square Saturn on Thursday gives you a lift as far as the Post Office.

Lucky stuffing: Horsehair
Lucky craft: Macrame

Libra

This week you will be in your element (which actually turns out to be Caesium) so you above all others will appreciate that it takes a strong person to admit when they're wrong. It takes an even stronger person to go up to the first person and suggest that they are the one who is mistaken, so make sure you hire one.

Lucky force: The British Transport Police
Lucky drop: Pear

Scorpio

Over indulgence in good wine has left you with a predisposition to break into the Macarena – despite the effect that it has on your knees. A recent holiday may not have proved as restful as it could have been due to the over-zealous attentions of security staff who confiscated your best 'Dora the Explorer' biro on the grounds that 'it could have someone's eye out'.

Lucky shoes: Slingback Brogues
Lucky aroma: Brilliantine

Sagittarius

A unique aspect between the Moon and Mercury indicates that your recent sinus trouble is being aggravated by exposure to a fungus found only on ancient Roman war-galleys. On Wednesday you may suffer catastrophic and sudden tie failure. Try not to over-compensate.

Lucky battery: Triple A
Lucky snack: Ptarmigan custard

Capricorn

A square Saturn can only mean that your hobby of rearranging your sock drawer is beginning to lose its thrill. On Wednesday as Venus comes into aspect you will be offered the role of Mr. Waverley in a remake of *The Man from Uncle*. It's probably best not to mention the corncrake incident at the audition.

Lucky dismount: Double pike somersault
Lucky seaweed: Bladderwrack

Aquarius

With Mercury, planet of communication, square to your birth-sign on Tuesday, you will receive notification of a forthcoming award. The Licensed Victuallers Association will announce that you are the winner of the Winehouse challenge trophy. As this will be your third such award, you will keep the cup. Beware of eggy soldiers on Friday.

Lucky pastime: Shouting
Lucky nails: Masonry

Pisces

The New Moon was in direct opposition to Pluto over the weekend, so it is unlikely that your offer of marriage to Des Lynam will be any more successful this time than for the last seven. An unusual aspect between Mercury and Saturn on Wednesday will mean that you will have a shifty look about you for the last part of the week.

Lucky supplement: Cod-liver oil and malt
Lucky plant: Turntable crane

Weekly Forecast for 25th April to 1st May

Aries

A quiet week during which the only unusual activities may be a call to act as a character witness for an erstwhile acquaintance. A rising Venus combined with a trine Neptune indicates a chance meeting at the 'dogs' in Walthamstow will lead to a potential bargain on a job-lot of shop-soiled heated towel rails.

Lucky horse: Vaulting
Lucky wilderness: Croydon

Taurus

If it is your birthday this week, Mercury square to Mars on Tuesday indicates a more conservative celebration than last year's debauched Bacchanalian brawl through every public-bar in Barking, Dagenham, and Upminster. This year you will give Upminster a miss. Later in the week, you will have the opportunity to enhance your standing at work. Just remember, if you cast your bread upon the water, you'll get really soggy bread.

Lucky clientele: Discerning
Lucky eye shadow: Taupe

Gemini

Love is in the air. On Tuesday, Venus enters your birthsign without knocking and you meet a tall shapely stranger. This will be your first romance with a basketball international so allow yourself to be a little more impetuous than usual. Why not try out that new pedestal set, and leave your flat shoes at home?

Lucky paste: Bloater
Lucky herb: Alpert

CANCER

Mars in Capricorn has been causing you to have a recurring dream in which you become involved in an unseemly brawl with the Royal family staff in the staff canteen. You always wake at the same moment. Just as you are about to push Prince Michael of Kent's head beneath the surface of the Brown Windsor for the third time, you are pulled off by Princess Anne. Jupiter in your fourth house indicates that this could be down to Cheese and pickled-egg toasted sandwiches for supper.

Lucky spider: Harvest
Lucky circumstances: Unforeseen

LEO

Last week's Lunar Eclipse in Taurus indicates that the week at work will be a typical triumph, i.e. noisy, oily, and won't start in the mornings. On a brighter note, Thursday brings good news that you are the direct descendant of the person who invented the circle, and are now due four million years of royalties.

Lucky gas: Helium
Lucky haircut: Short bob

VIRGO

Mars rising on Tuesday will mean that the recently installed security scanner that reads your palm-print on entry to the building tells you that you will meet a tall dark stranger who once bought a Morris Marina from the Duke of Westminster. This will prove to be false information caused by a fault in the software. It was in fact a Morris Oxford. Fortunately when you cross the palm-print with silver, it lets you in anyway.

Lucky topping: Royal icing
Lucky seabird: Cormorant

Libra

A trine Mars at the weekend means that you may have put on a little extra weight in the last few days. You should not be too concerned that you now stall airport escalators, as fickle Pluto enters your birthsign on Monday bringing with it a fierce case of 'Bosphorus Phosphorous'. By Thursday you should be on the mend and able to remove the Andrex from the freezer.

Lucky language: Bantu
Lucky tattoo: Edinburgh

Scorpio

Venus rising at the beginning of the week indicates you'll fall foul of the Ministry of Agriculture Fisheries and Food when they discover your plan to take up whaling on a commercial basis. Saturn entering your fourth house on Wednesday will mean a possible brush with the law when you're caught making unseemly gestures at the builders across the road from your office window.

Lucky habit: Snuff
Lucky drink: Lager-tops

Sagittarius

The New Moon in Taurus indicates a tough week for anyone closely involved with speed skating. The effects of last week's eclipse are still evident in your birthsign, particularly at the end of the week when you are plagued by Sagittarius rising which will wake you in the early hours of Friday by slamming the car door, and falling through your hedge.

Lucky order: The Carmelites
Lucky soup: Cock-A-Leekie

Capricorn

Fortune smiles on the aristocrats of the zodiac this week, but your decision to attempt to grow the world's longest handlebar moustache is thwarted by an unlucky combination of a pair of pinking shears and an apprentice dervish you accidentally brought home in your hand-luggage from a recent business trip.

Lucky fish: The Grayling
Lucky poultice: Mustard

Aquarius

Venus transiting the Sun may manifest its influence in a number of unusual ways this week. On Wednesday while slicing Aubergines for your act, you discover a pattern of seeds that looks exactly like Alvin Schockemohle. On Friday, an adverse reaction between your trousers and shirt will cause a quiet tweeting noise every time you try to use an adverb.

Lucky garnish: Chives
Lucky teeth: Incisors

Pisces

Pluto and Neptune conjoin in your birthsign this week leading to disharmony and discord. This is most evident in a choral recital you give at the end of the week, which even your best friends will describe as sounding like an angry man clubbing a muskox to death with a set of bagpipes.

Lucky rhythm: Syncopated
Lucky starch: The plantain

Weekly Forecast for 2nd to 8th May

Aries

The link between Venus and Jupiter on Thursday indicates that you will need to start thinking about a new way of dealing with young people. The routine of pretending to steal their nose rarely works with anyone over three years old. On Wednesday you will have a run in with the dry cleaners over frequent zip repairs.

Lucky adjective: Sprightly
Lucky muse: Terpsichore

Taurus

A well-rounded look to the week as Venus and Mercury both aspect Pluto. You really need to consider how hard you've been working of late and think about giving yourself some well-earned time off. On Friday, you accidentally book a working-class return to Reading which takes the whole weekend to recover from.

Lucky footwear: Suede brogues
Lucky glaze: Gelatine

Gemini

Venus transits the Sun on Tuesday, which indicates an unusual opportunity to turn a long cherished dream into reality this summer. However, mischievous Pluto rising means that you should confirm arrangements with the travel agent in writing rather than on the phone, unless swimming with moleskins is your idea of a good time.

Lucky fritter: Banana
Lucky cup: 34C

Cancer

If you start the week with financial worries there is a good chance that these will have eased by Friday when you accidentally come up with an unbeatable idea while paving over your window box. An unlikely sequence of events sees you combining a recording of your trademark giggle, a hover mower, and some red nylon fur, thereby inventing the 'Tickle Me Flymo'.

Lucky garnish: Knob of butter
Lucky mollusc: Snail

Leo

Your Venus is looking increasingly ill-at-ease of late which may help to explain the trouble you've had getting satisfactory results from your new trouser-press. However, the New Moon in Capricorn will see the dreams about Ken Livingstone tailing off during the course of the next week or so, and a warm front moving in from the West.

Lucky pudding: Tapioca
Lucky table: Six-times

Virgo

Saturn's benign influence will bring joy and happiness to everyone who realises that they are not as old as you. Venus rising means that you may be approached in the street by a strangely attractive woman who insists on a birthday kiss, passing on her address, and a mild but persistent gum disease.

Lucky vitamin: C Minor
Lucky hold: The folding trouser press

Libra

Monday's aspect between Venus and Neptune will mean a misunderstanding in the ten items or less queue at the double-glazing showroom involving three French hens, a knitting machine, and The Shadow Chancellor of the Exchequer – again. A dramatic trine between Mercury and Mars on Tuesday indicates a Bacchanalian lunch with a medical man. So don't wear your best trousers – you know what you found in the turn-ups last time.

Lucky accent: Somerset
Lucky Starr: Edwin

Scorpio

An unusual aspect between Mars and Neptune has kept things fairly quiet for a week or so, but all that will change on Wednesday when you are given a pair of working chameleon-skin trousers that enable you to blend into any background from the waist down. These will turn out to be particularly useful on Thursday when senior managers decide on a snap cubicle inspection.

Lucky road-kill: Hare
Lucky particle: The quark

Sagittarius

This week your ruler the Sun is trine with Jupiter which means some respite from the constant battle to wring the best out of the work-shy fops with whom you are burdened. Look on the bright side, you have an attractive mower, a wife who starts on the third or fourth pull, and a world-class collection of reading glasses, so things could be much worse – and indeed on Friday, they are.

Lucky loins: Pork
Lucky alias: Thaxted Mulrooney

Capricorn

Over the years, your quick and clever tongue has made you popular on the whole, and it has surprised many that you have not risen further. However, on Wednesday, the Full Moon out of your office window will have a dramatic effect on your fortunes when you are spotted by a senior television executive, who immediately signs you up as Ant & Dec's stunt double.

Lucky snack: Bread-pudding
Lucky Parton: Dolly

Aquarius

Mysterious Neptune enters your seventh house on Wednesday bringing with it a sense of foreboding along with an unexpected bill for emergency chiropody work. Many people see you as being a little timorous, shy, and even insecure. But be of good heart, with a little more effort, the restraint you inherited from your Mother should finally give way after one more really determined chew on Wednesday night.

Lucky beetle: Goliath
Lucky paper: The Essex Chronicle

Pisces

Although the years have so far been kind to you, fickle Pluto is a constant reminder that the constant strain of early mornings and early lunches may one day begin to take its toll. Saturn is trine with Neptune on Thursday which indicates that you may pick up a touch of St. Elmo's fire, so keep a bucket of water about you at all times.

Lucky composer: Schubert
Lucky symbol: Ampersand

Weekly Forecast for 9th to 15th May

Aries

You have always enjoyed the finer things in life, but your ruler, Saturn, entering your fourth house on Tuesday, combined with a surprise phone call from NASA telling you that you are starting to show up on satellite photographs, indicates that cutting back on the nourishment may be no bad thing. Fate however takes a hand on Thursday when you lose nearly three stone in the back of a taxi in a freak wonderbra accident.

Lucky phobia: Protozoa
Lucky hold: Half-Nelson

Taurus

An unusual start to the week is indicated by Mars rising. It seems that after years of pulling faces to amuse and entertain senior executives, your Mother's warning finally comes true, the wind changes (no bad thing in your case) and you stay that way. Fortunately, a trine Mercury means that the resulting rictus will only last until Thursday, when a freak paper-cut severs the nerves causing the problem.

Lucky tern: Sooty
Lucky Pope: Pius X

Gemini

Venus, planet of love, moves into your sign on Wednesday indicating that you will meet the woman of your dreams. Unfortunately, as usual, you will be sound asleep at the time. Later in the week, Jupiter forms an unusual square with Pluto and Neptune leaving you unable to shake off the eerie feeling that you've had déjà vu before.

Lucky tone: Wolfe
Lucky triangle: Winnersh

Cancer

A trine between dynamic Mercury and your ruler, the Sun, means that you need to take care not to become exhausted by your heavy workload. Now might be the time to consider a break – a line-dancing weekend in Bratislava, or even that cross-country double-entry book-keeping holiday in Rhyl you've had your eye on.

Lucky estuary: Medway
Lucky accent: Flemish

Leo

Venus is in your birthsign this week, so the prospects of a real old-fashioned romance are good, particularly on Monday. Even if you don't find the love of your life this week, you will certainly get a surprise on Tuesday, as Mars in your fourth house indicates that you will be ambushed and carried off by hover-flies when you wear your second-best yellow shirt to work.

Lucky process: Photosynthesis
Lucky sidekick: Dr. Watson

Virgo

Mercury, planet of communication enters your sign on Tuesday. This should see an end to an ongoing disagreement you have had with the HR department over your interpretation of the staff disciplinary procedures. Your suggestion that these should include Chinese burns, horse-bites, nutmegs, and for lesser offences, a game of 'knuckles' with the head of Marketing, is not well received.

Lucky estuary: Medway
Lucky main course: Boiled chicken nerves

Libra

Everyone will notice Libra this week. You're the ones with the glow around you that nothing can dampen. Later in the week, Mars rising indicates that you will discover that this is due to an excess of seafood in your diet, which has caused you to develop a phosphorescent 'wake'.

Lucky vein: Vena cava
Lucky harp: Aeolian

Scorpio

On Tuesday, the New Moon in Pisces indicates that your true talent will at last be recognised and a big Hollywood film studio will offer you the part that you were born to play, that of the leading role in the autobiographical story of Mick McManus. On Friday you will have ham and boiled new potatoes for dinner.

Lucky spasm: Restless leg
Lucky sauce: Hollandaise

Sagittarius

Saturn rising in your third house indicates a strong interest in the activities of your close neighbours, especially the lady across the road. On Thursday mischievous Pluto enters your birthsign which means that you may well forget to take down the tripod and telescope before drawing the curtains to leave for work. The good news is that your photography is improving no end.

Lucky cuffs: French
Lucky dance: Gavotte

Capricorn

Venus and Saturn conjoin in your birthsign this week bringing an opportunity to control the volatility that has characterised so many of your recent pie recipes. On Wednesday a freak accident with a bulldog-clip and a tin of mustard powder will leave you with a long-term inability to swallow wine-gums.

Lucky tense: Past participle
Lucky tipple: Sloe gin

Aquarius

This week's Full Moon suggests that you will discover an impressive and potentially lucrative capacity for mashed potato consumption. However, there could be trouble in store at the end of the week when a retrograde Pluto means that a close personal friend may have a mishap which results in them spilling blanc-mange on your Autoharp.

Lucky curtains: Flameproof
Lucky language: Norwegian

Pisces

The Sun's sharp link to Venus on Tuesday indicates that you will be swept away on a roller coaster of powerful and conflicting emotions, veering between yearning passion, and quiet contentment when you receive some unsolicited slippers through the post. The Full Moon will also mean a touch of indigestion on Monday afternoon just before Tiffin.

Lucky precaution: Shillelagh
Lucky wavelength: 20 kilohertz

Weekly Forecast for 16th to 22nd May

Aries

You are feeling a little fragile at the beginning of the week following your party on Saturday to launch your latest invention – the Crosse & Blackwell your heart, Mulligatawny Bra. On Thursday Pluto rising indicates that you may be tempted to buy into a syndicate involved in the import of false nose-hair. Resist it with every fibre of your being.

Lucky spasm: Dry heaves
Lucky fencing: Woven hurdles

Taurus

On Monday mysterious Neptune conjoins with Mars in Capricorn. This can only mean that you will be afflicted with a mysterious but mild ailment that baffles your Doctor. Towards the end of the week, a trine Saturn means that the medical authorities finally find out what is wrong with you just days too late to have the affliction named after you.

Lucky rodent: Coypu
Lucky choking hazard: Poker dice

Gemini

An interesting and enlightening week ahead. On Wednesday, a friendly aspect between Venus and Mars means that you will make an important discovery. You will find that when pensively forming your fingers into a 'steeple' during important meetings you will become much more credible when you don't progress to the next stage and wiggle all the 'people' at those present.

Lucky jigsaw: Durham Cathedral
Lucky tree: The Larch

Cancer

Although you don't normally have difficulty expressing yourself, Monday will see you suffer writers' block when you are making up the monthly management reports. An uncomfortable trine between your ruler, the sun, and Jupiter indicates that on Thursday lunchtime you will be chased the length of the High Street by a man dressed as a crested-grebe and armed with a selection of Deryck Guyler memorabilia.

Lucky mood: Upbeat
Lucky moustache: Handlebar

Leo

Now that your ruler, Mercury, is back on its direct course and moving back into Virgo, you will notice a quickening in your pace although this may have more to do with the double enchilada platter you had for Sunday lunch. On Tuesday Uranus, planet of unexpected developments enters your birthsign so a fierce looking woman, adept in several languages may make a covert attempt to win your affections at line-dancing class.

Lucky carnivore: Wolverine
Lucky paste: Wallpaper

Virgo

Jupiter, the giver of gifts and luck, will leave your fourth house on Wednesday, taking with it some of the ornaments you've been meaning to give to the local charity shop. On Friday, Venus rising indicates that a chance meeting in the local abattoir will lead directly to an unsolicited offer to alter your bedroom curtains.

Lucky sauce: Worcester
Luck stone: Kidney

LIBRA

Despite a few days well-earned rest in Walton-on-the-Naze, you may have a tendency to get a bit irritable this week under the influence of unpredictable Pluto. On Thursday, an unusual aspect between the Sun and Mars means that you will be seized by an overwhelming urge to shave your kneecaps in the bath.

Lucky custard: Strawberry
Lucky gas: Hydrogen

SCORPIO

This week, Mars rising indicates a somewhat depressing week full of self-doubt, health worries, bitterness and past embarrassments, will not trouble you even slightly. However on Friday you will accidentally guess the secret of an isolated rural community and be justifiably ostracised.

Lucky vessel: Cutter
Lucky sensation: Prickly heat

SAGITTARIUS

As the period of change that began with the New Moon in Libra moves into a new stage, you will be less troubled by persistent spots than in recent months although blocked pores may still prove problematic. Don't be tempted to visit the all-night Milliner until Mercury goes into trine on Friday.

Lucky cartoon: Scooby-Doo
Lucky curry: Prawn mufti

Capricorn

Your drive and ambition is no secret to anyone who really knows you. On Thursday Venus enters Virgo indicating that you will get a unique opportunity to further enhance your status by acquiring a set of personalised dental plates. An unusual aspect between Mercury and Saturn on Friday means that you may need to be more vigilant than usual for renegade Commanche raiding parties in the High Street.

Lucky snack: Salted cashews
Lucky lake: Superior

Aquarius

A quiet week, during which you should start to plan for your forthcoming concert tour. Mars in Aquarius suggests that you would do well to keep on the right side of your road-crew. The correct term for these people is 'Roadies' rather than 'Toadies' as you have been calling them. This may go some way towards explaining the lioness in your sock drawer this morning.

Lucky confection: Sherbet-dab
Lucky trousers: Blue corduroy

Pisces

A hitherto undiscovered inventive streak comes to the fore on Tuesday when you broker a deal with the high-street banks to produce the ideal gift for your older friends and relatives – an 'in my day' gift token – valid for a pint of milk, a pat of butter, a loaf of bread, an all-day gobstopper, a trip to the pictures and change enough for the tram ride home.

Lucky platitude: Some people, eh?
Lucky infestation: Red spider mite

Weekly Forecast for 23rd to 29th May

Aries

Neptune rising on Monday indicates that at around teatime this evening, you will have someone push a ripe banana through your letterbox bearing the legend 'Remember Bratislava?' in green ink. Do not eat the banana, as green ink is bad for you. On Wednesday, in return for your hand in marriage, a man dressed as 'Jelly-Roll Morton' will give you a delicious recipe for a 'Whippersnapper Glory'.

Lucky fuse: 3amp
Lucky pattern: Paisley

Taurus

Mars is square to Mercury in your fourth house at the beginning of the week. On Thursday lunchtime, you will become involved in an unseemly tussle over the last sausage-roll in the canteen with 'Rollerball Rocco' and 'Skull Murphy'. The good news is that after two grapevines, a 'Boston crab', and a Corby trouser press, you win by a fall and a submission. Mind your back.

Lucky Partridge: Don
Lucky fungus: Mildew

Gemini

An incident-filled week with – and this is no surprise to those that know you – the emphasis on high fashion. But trendsetting can have its drawbacks. Your recent penchant for mixing plaids and tartans of differing hue and pattern will get you into hot water on Wednesday when a trine Pluto in your seventh house indicates that you will be arrested in Wandsworth for 'aggravated Burberry'.

Lucky arrangement: By mutual consent
Lucky settlement: Out of court

Cancer

After a quiet start, this week looks to be one of the dullest in your chart. The confluence between Mars and Neptune last weekend means that while shopping for new plimsolls at the local market you will bump into Hollywood film star Kiefer Sutherland. However, the malign influence of Pluto means that you will annoy him by asking if he has an older and younger brother called 'Kief' and 'Kiefest'.

Lucky punch: Right hook
Lucky mating call: The giant Pacific clam

Leo

Saturn transits your birthsign on Tuesday. This, combined with an unusual aspect between the Moon and Neptune indicates a chance meeting in the Co-op with a manufacturer of novelty reading glasses from Kempten in Bavaria. Don't waste your time asking if his local racetrack is five miles long (Oh, di-doo-da-day), as he will have heard that one before.

Lucky cough: Dry tickle
Lucky odds: 11/2

Virgo

In recent months you have been tied down much more than you normally enjoy. And if you find yourself strapped for cash later in the week, you only have yourself to blame. Your ruler, Mercury, is going to be travelling in tandem with Venus for the next week, so make sure you use plenty of talcum powder and don't forget your stretching exercises – tendons don't grow on trees.

Lucky tone: Conciliatory
Lucky manner: Bedside

Libra

Always a worrier, you may be too hard on yourself just now. No-one could have foreseen that you would be undercut by Phil Collins when applying for your usual part-time Santa job in Selfridges. On Friday, when Saturn and Pluto are once more in exact opposition, a man in a bootlace tie with a Hornbill under one arm will give you a funny look in the Public bar of the Alma.

Lucky tempest: Troy
Lucky chair: Hepplewhite

Scorpio

You will have no trouble standing up for your point of view, but exercise diplomacy, take a deep breath and count to ten if there are arguments with your partner on Wednesday, as they will be in the mood to flay the skin off you. Pluto travelling through your fourth house on Tuesday means that you might fancy a slice of Battenberg cake with your usual leek soup for elevenses.

Lucky veneer: Birch
Lucky alloy: Pewter

Sagittarius

Although Mercury and Venus are travelling together in your sign, providing a congenial backdrop to your life, each planet has an awkward aspect to Neptune early in the week, which means that you will temporarily lose the ability to play the tenor-saxophone. Your fan mail reaches a new high on Thursday when you receive over 13 letters proving conclusively that beauty is in the eye of the bewildered.

Lucky position: Precarious
Lucky bat: Pippistrelle

Capricorn

Saturn transits your fourth house on Monday providing an ideal opportunity to really clean up, and indeed shine when you park your car under a tree much favoured by pigeons. A drastic shortage of soft tissue in the staff facilities at work is indicated by the rare conjunction between the Sun and Uranus. On Friday a square Pluto means that you will feel kindly disposed toward hedgehogs all day.

Lucky wolf: Maned
Lucky iron: 7

Aquarius

The malign influence of mischievous Pluto on Tuesday means that this will not be a good week for studying forestry. A rare opportunity to appear in a documentary about commercial pancake mix will nearly be missed on Friday lunchtime when you are delayed by an unusually dense fog in the corridor on the first floor.

Lucky bank: Dogger
Lucky hinge: Rising Butt

Pisces

Saturn's square with the Sun is urging you, and anyone else who shares yours, or indeed any other birthsign, to seriously consider posting poison spiders to the person who thought up 'Wife Swap'. On Friday, a large man with a big nose and a rolled-up copy of *The Sporting Life* may lunge at you with a set of laundry tongs. Remember your training.

Lucky clip: Bulldog
Lucky pub: The Royal Steamer

Weekly Forecast for 30th May to 5th June

Aries

Mars entered your birthsign on Sunday, and with it the indication that this week your dreams could hold a message for you, especially the one where you're floating in a pool full of Swarfega and giant ants, each with the face of a well-known cricket commentator, that swim up to you and ask where they can buy a decent hot-water bottle. You might like to consider cutting down on the cheese and biscuits for supper.

Lucky laminate: Formica
Lucky village: Meldreth

Taurus

Mars enters your fourth house on Wednesday morning increasing the likelihood that you may be accidentally captured in a series of bucolic landscape paintings with a selection of hooved ungulates On Thursday you will finally see a breakthrough in your quest for the perfect Birthday present for your partner when you spot a travelling executive wormery on the shopping channel.

Lucky dressing: Ranch
Lucky vitamin: D Major

Gemini

An unsettling week during which Venus transits your birthsign. This combined with the waning moon in Aries indicates that you may be mistaken for a former secret agent and sent a pair of poisoned shoe-trees in the internal post. On Friday your ruler Mercury moves into dynamic Scorpio which signals a welcome change from the ginger nuts you've had to endure for elevenses for the last few weeks.

Lucky live-bait: Ragworm
Lucky festival: Harvest

Cancer

Be careful when doing anything to excess, especially on Wednesday, and doubly so if it involves tapioca. Wasting money on items that appeal to your fancy now but do not have any lasting merit or appeal is particularly enjoyable at the moment. The one exception being the new range of 'forbidden fruit of the tree of knowledge' Jam, which is particularly good on wholemeal toast.

Lucky soup: Cream of asparagus
Lucky foible: Pencil erasers

Leo

This is a very exuberant time, with high energies. You feel physically strong and fit, and you are even more willing than usual to become involved in re-enactments of famous battles of the Punic wars. Neptune is conjunct to your seventh house on Tuesday which means that you stand a better than average chance of being struck by lightning in the hairdressers.

Lucky lawn: Chamomile
Lucky pudding: Dundee rice

Virgo

Last night's Full Moon, sextile with Venus means that you may have some issues with colleagues this week. Try to think of those you work with as pieces of furniture, and those in your department as the legs on a table. One leg is respect, one is trust, one is shared pleasures, and one is shared dreams. On a lighter note, Mars rising means that someone will pop along with some beeswax and new dusters on Thursday.

Lucky frame: Roddy
Lucky mood: Playful

Libra

Uranus, planet of surprises, has plans for you this week. However, these may have to be put on hold as a trine Pluto indicates that on Tuesday lunchtime, while at lunch with a colleague, an undercooked turkey may cause you problems when you sit at the restaurant table under which it has been hiding. On Thursday you may be arrested for unlicensed haberdashery in Market Street.

Lucky pantomime: Jack & the Beanstalk
Lucky muscles: Trapezoids

Scorpio

The relentless quest for the answer to the age-old problem of preventing peas going cold before everything else on the plate has occupied your waking life, but this week, volatile Neptune enters your birthsign bringing with it the inspiration to start a bumble-bee farm in order to make individual pea-cosies from their furry pelts. On Thursday, you will be roughly jostled in the bargain basement while searching for a Christmas present.

Lucky paper: The Daily Sketch
Lucky sect: The Rosicrucians

Sagittarius

An unusual aspect between Saturn and Mercury indicates that you may be drawn into fruitless shopping trips over the next week or two, or at least until Mars goes trine on Friday week. Resist any suggestion that might be made that your carrying capacity would be dramatically improved were you to be fitted with panniers and a tow-bar. On Wednesday the old trouble will flare up again. Pretend not to notice if people stare.

Lucky snack: Spaghetti-hoops 'a la Grecque'
Lucky guess: Mandibles

Capricorn

An action packed start to the week is indicated by Mars conjoining with Pluto in your seventh house. In a vain attempt to increase your 'street cred' you don an elaborate disguise and get a job reading the news on a commercial radio youth station. But a square Saturn means that this will be a short-lived venture as during the first bulletin you are unmasked as you correctly conjugate a verb.

Lucky cake: Eccles
Lucky shoes: Winkle-pickers

Aquarius

Your deep affection for risqué verse may have serious consequences on Wednesday when in an unguarded moment you provide an example to someone you meet in the corridor, who actually turns out to be a well-known Bishop whose hobbies include keeping owls and making his own pickled eggs. On Friday mysterious Neptune enters your birthsign, which is a bit of a nuisance.

Lucky spaniel: King Charles
Lucky tool: Spokeshave

Pisces

Mercury goes retrograde in your birthsign on Monday this week. This combined with a waning moon indicates that a forthright woman with a pronounced lisp will brush against you in the lift. Not until Tuesday lunchtime will you learn that you're 'it' and no returns.

Lucky pipe: Meerschaum
Lucky hold: Stratford lift

Weekly Forecast for 6th to 12th June

Aries

You know that feeling of holding your breath before something happens? Well, just for a change the feeling is correct – something really is just around the corner waiting for you now. Saturn entering your birthsign indicates that all the washing up from the weekend has been left in the sink, and the plughole is blocked with soggy tea bags.

Lucky rug: Shepherd-skin
Lucky stranger: Steve Corbet

Taurus

Mercury is trine with mysterious Neptune this week, which means for you that a certain rhyme and rhythm will be associated with everything you do. This will be particularly noticeable on Wednesday lunchtime when you purchase a harmonica at a music shop run by Monica and Veronica Honniker. However, you should not attempt to push your luck by asking if they'd care for afternoon tea.

Lucky walk: Cheyne
Lucky compost: John Innes No. 1.

Gemini

Last week things were changing so swiftly that you had to spend a lot of time on your toes, making decisions from moment to moment. Unfortunately your toes are not what they were, and Mars in Aquarius suggests that foot-care products and chiropody could be paramount in your thoughts in the coming week.

Lucky dip: Avocado and walnut
Lucky gravy: Onion

Cancer

On Monday, your ruler, Mars, is in favourable aspect to Pluto. This is likely to indicate a wind of change blowing through your chart. Sure enough, on Tuesday lunchtime you will receive a long overdue initiation to appear on *Pro-Celebrity One Man and His Dog*. Bring a whistle.

Lucky exclamation: Spiffo!
Lucky snack: Blinnies

Leo

Scientists have recently discovered that if you feed one aspirin a day to laboratory rats, eventually you are going to get bitten. This sums up your week. An unfortunate trine between Pluto and Mars indicates that you will be attacked and mauled by scientists justifiably angered by your overbearing bonhomie during vespers.

Lucky drum: Oil
Lucky alloy: Pewter

Virgo

You've been a bit worried whether you might be losing your grip recently, but Saturn's powerful influence enters your birth-sign on Tuesday, immediately providing a positive feel in the lift on the way to the staff canteen. Fortunately, as the lift will be crowded at the time, nothing will be attributable to you. Try to cultivate an innocent look just in case.

Lucky Handl: Ireni
Lucky shoes: Winklepickers

Libra

The Sun, your ruler, in your opposite sign of Aquarius, is trine Saturn, which indicates that you should make sure you don't run out of nasal spray on Wednesday. If you spot any particularly cheap lino in town, buy it. Before the week is out you will have the chance to swap it for some show-cavies.

Lucky tipple: Snowball
Lucky game: Pheasant

Scorpio

Stubborn Pluto in your sign at the beginning of the week means that you may have trouble selecting a suitable birthday card for your beau. Although hearts are usually a popular theme, and even though you may wish to convey your true feelings, a stuffed heart from the local butcher may not achieve the results you were expecting.

Lucky wallpaper: Flock
Lucky rub: Dutch

Sagittarius

You are, as most people know, a person of many parts, however, Saturn conjoining with Jupiter in your seventh house indicates that your secret identity will not remain so for much longer. The Protractor – a shadowy figure who stalks the streets at night hunting down people who can't do geometry, and sticking compass points in their bottoms – will finally be unmasked.

Lucky place: Peyton
Lucky clippers: Tea

Capricorn

On Monday a trine Mercury means that you will meet a lot of people with money, none of which is coming your way, however, Venus enters Aquarius on Tuesday and with it comes the chance you've been waiting for to break into modelling. You are to be used as the model for the new economy size 'Matey' bubble-bath pack (25% extra). See if you can get your mum to dig out your old sea-scouts uniform.

Lucky film: 35mm
Lucky exposure: Indecent

Aquarius

A busy week is indicated by the arrival of mysterious Neptune in your birthsign. On Monday you'll invent a party game for people who really don't like party games, although the title, 'well, if that's what everyone else wants to do', may need some work. On Thursday you may be lunged at in the Library by a lady who looks like Joan Hickson.

Lucky moustache: Pencil
Lucky gland: Thyroid

Pisces

Any kind of activity involving spirituality, mysticism or omelettes will be very good for you, except on Monday, when any advice you receive and give will be utterly worthless. A square Saturn on Tuesday will leave you regretting your rediscovery of Port shandy yesterday evening.

Lucky ointment: Germolene
Lucky plum: Victoria

Weekly Forecast for 13th June to 19th June

Aries

Your recent holiday has done much to recharge your batteries. Venus rising will see you in the midst of things socially, but a misunderstanding about the correct plural for Weetabix will lead to a bit of a scene at the slipper baths on Tuesday afternoon. Warm up properly.

Lucky eclipse: Lunar
Lucky vitamin: C Major 7th

Taurus

You've really got the bit between your teeth this week. With the Sun winging its way through Leo you'll think you're unstoppable. Make the most of it, as a chance encounter with the Deputy Prime Minister on Wednesday will cause a mood swing and you'll spend two days eating nothing but trifle.

Lucky plant: JCB
Lucky bunting: Corn

Gemini

Love is in the air. On Tuesday Venus enters Sagittarius and you meet a tall dark stranger. This will be your first romance with a stilt walker so allow yourself to be a little more impetuous than usual and leave your flat shoes at home. On Thursday, a brewing disagreement among peers comes to an ugly head. And though you know you should stay out of it, your years as a cage-fighter stand you in good stead.

Lucky vegetable: Parsnip
Lucky herb: Marjoram

Cancer

Mars entering your fourth house on Tuesday means that you will be even more in demand than usual. Wednesday will see you offered a speaking engagement at the International Aroma Producers Convention in Basildon. An impulse buy of smoked haddock on Friday may lead to you meeting the woman of your dreams on the fish counter. Unfortunately your dreams often involve an 18-stone tattooed lady with dyed green hair.

Lucky scale: Beaufort
Lucky alloy: Brass

Leo

This week finds you descending into a cleaning fervour, during which you discover rare and interesting personal antiquities. Maybe it's an old photograph, or a love letter, or that evidence acquitting your Aunt May of involvement in the great Train Robbery. Avoid Guildford for the rest of the week.

Lucky packaging: Bubble-wrap
Lucky laces: Leather

Virgo

A chance meeting with a trichologist in Reading on Wednesday lunchtime will finally reveal that you have been buying the wrong type of shampoo all these years, and should have been investing in care products for 'angry hair'. Take extra care if sitting on unfamiliar furniture on Friday as Pluto rising indicates that a combination of shoddy workmanship and poor quality glue may lead to an accident.

Lucky solvent: Acetone
Lucky movement: Pincer

Libra

Cheese will feature heavily in your chart over the coming week, and may well begin to appear as a part of elevenses once again. Your ruler, the Sun, transits Neptune on Thursday and sees you among friends and admirers – perhaps even setting up a new Political party.

Lucky game: Mousetrap
Lucky pancake: Crispy duck

Scorpio

Your ruler, the Sun, finds itself square Mars yet trine mysterious

Neptune. All this means that with Mars ascendant you can really start to act like the king of the jungle you are. Although raw meat may appeal this week you should resist the temptation until later in the month. Meanwhile, make do with a research assistant, or at a push, an egg and cress sandwich.

Lucky smell: Creosote
Lucky mammal: Mongoose

Sagittarius

Venus transiting the Sun may manifest its influence in a number of unusual ways this week. On Wednesday while shopping for new equipment for your act, you meet a well-dressed, but extremely hairy woman who offers you a walk on part in the Hollywood remake of *Wagoners Walk*.

Lucky garnish: Angelica
Lucky teeth: Top-set

Capricorn

You may have wondered why the scaffolding had not been removed from the building opposite. The explanation is the obvious one. As Pluto is entering your fourth house, and the Moon forms a trine with Neptune, this indicates that you will wake tomorrow morning to the sight of John Prescott dressed only in a short-kimono bathrobe doing 'hand-stands' on the front lawn.

Lucky pie: Apple and blackcurrant
Lucky sensation: Goose pimples

Aquarius

Your passion for bath-toys has been an open secret for some years. A square Venus at the beginning of the week indicates that it's high time that you owned up to having become the largest wholesale supplier of battery operated flotation devices in the southeast. On Tuesday, the return of Pixies downstairs indirectly influences your ability to get a decent quote for pet insurance. There may be grounds to sue.

Lucky pudding: Gypsy tart
Lucky establishment: The soapy frog

Pisces

Change is in the air, which of course means that the furniture will be on the move again, and there will be much talk about redecoration. When it comes to selecting wallpaper and paint, you have no peer. Saturn goes trine on Thursday, so if you do have particular hues in mind, be sure to make this clear, otherwise the ever popular Grant and Fearnley-Whittingstall may be proposed.

Lucky sealant: Unibond
Lucky bathroom suite: Avocado

Weekly Forecast for 20th to 26th June

Aries

Once you would have rejected outright the kinds of offers that you are now considering, but you are right to be looking at some of the unusual options coming your way, and with Wednesday's lovely aspect between Saturn and Neptune, you would be ill advised to pass up the opportunity to host the latest TV reality cooking show, *Pets Make Pies*, which pits two families against each other in a trivia quiz, with hilarious pastry-based consequences for their cherished companion if they lose.

Lucky current: Humboldt

Lucky punch line: That's okay, I'm not a real welder.

Taurus

Mercury is square to Mars this week, so any respite from the problems you've been having with your continental quilts is unlikely. On Wednesday you wake to the sound of loud tutting outside your front door. On investigation, this proves to be an impromptu rehearsal by yet another Skippy the Bush Kangaroo tribute act.

Lucky wrestler: Jackie Pallo

Lucky polish: Cardinal Red

Gemini

Saturn enters your seventh house on Thursday and leaves a stain on the stair carpet. Despite your early misgivings, Venus is spreading a warm influence in the area of your chart concerned with infantry manoeuvres, so it shouldn't come as too much of a surprise to find the Royal Fusiliers in your Forsythia again on Thursday. Try spraying with a mild solution of Milton fluid.

Lucky adjective: Creaky

Lucky jam: The M40

Cancer

The influence of your ruler, the Moon may mean that you feel a bit quiet and thoughtful this week. Next week holds out the prospect of something very romantic, and at least one of your dreams coming true, though not necessarily the one in which you are invited to share an after-match bath with London Irish.

On Thursday an arm-wrestling match with Moira Stewart may not go as planned.

Lucky lotion: Camomile
Lucky pencil: HB

Leo

On Tuesday, an interesting aspect between Venus, Pluto and the South Node of the Moon means that a chance conversation between your next-door neighbour and a leading political analyst may mean that you end up taking both umbrage and Uxbridge at the general election.

Lucky sorbet: Kumquat
Lucky flooring: Cork tiles

Virgo

You're in for an emotional roller coaster this week. Mercury turning retrograde means that even though you are normally so practical and sensible, your hidden under-belly has always been sensitive and on Wednesday afternoon you discover that it is pursuing a solo career as a backdrop in the remake of *Quatermass and the Pit*. On Friday, a trine Pluto indicates that a missing hand brush may turn up in an unexpected place.

Lucky pump: Running
Lucky gland: Pituitary

Libra

The sheer energy from so many planets concentrated in Gemini this week means that you could find yourself in the grip of a strong but confusing emotional experience involving one or more of the cast or production crew of a West End musical, and a jar of crunchy peanut butter. With Mars moving into your birthsign on Thursday, extra Marmite might be a wise precaution.

Lucky Duck: Donald
Lucky Drake: Charlie

Scorpio

The energies are so strong and varied for anyone with a Sun sign this week. This, combined with Saturn rising could mean that by Tuesday morning you may well develop spectacular orange and green wings. Be especially careful with those around you on Wednesday. It's a day when you could cause hurt to a close friend with a poorly aimed Cornish pasty.

Lucky garnish: Chopped parsley
Lucky varnish: Nail

Sagittarius

By and large your sign has more energy than most, which is probably no bad thing because on Monday afternoon you are given 24 hours to find out exactly what 'by and large' means. Tuesday's New Moon provides a fresh perspective on an old problem when you are once again mistaken for Melvin Hayes while leafing through carpet samples in Cheesemans of Lewisham. Try to laugh it off without breaking into 'Meet the gang 'cos the boys are here'.

Lucky disguise: Horace Cutler
Lucky gesture: Dismissive wave

Capricorn

Romance may come suddenly into your life, in a new and surprising way when you are pounced on by a group of Korean gymnasts in your local shoe repair and key cutting centre.

Fortunately, their affections are short-lived when they discover that you are not the proprietor, merely a 'regular'. Mars rising indicates that a short man in very long trousers will try to sell you an invisible dog training-whistle on Thursday morning. Don't be taken in.

Lucky relation: Uncle Des
Lucky soup: Cream of pig

Aquarius

A chance remark overheard in the queue in Argos this week while picking up a new suitcase could mean that you become one of the country's leading saliva donors. Your predisposition for dribbling is well known, so why not cash in? On Friday, a badly placed 'chicken Fatwa' may take you suddenly in the Urals. Try not to flinch at the bus stop.

Lucky hold: Grapevine
Lucky floor: Parquet

Pisces

Mars in your opposite sign on Monday proceeds towards a hapless encounter with Uranus, which indicates that the authorities have noticed that the songbird population in your area has dwindled to perilously low numbers. On a brighter note, by the end of the week you will receive confirmation that your leopard-skin car seat covers are finally in stock.

Lucky sauce: Tomato
Lucky dwarf: Bashful

Weekly Forecast for 27th June to 3rd July

Aries

Jupiter is still retrograde in your sign. Its presence there is reassuring, but with it comes the ever-present danger that you will suffer a catastrophic dress-sense failure after attending one too many boot-sales. Do remember to read the forecast for your Moon and Ascendant signs before signing any electrical maintenance contracts.

Lucky accent: Mid-Atlantic
Lucky state: Rhode Island

Taurus

Oil reserves in Equatorial Guinea continue to be a worry for the remainder of this week. But at least Wednesday's Jupiter rising may go some way toward helping you unravel the meaning of your recurring dream about puff-pastry and the Beverley Sisters.

Lucky weakness: Port and lemon
Lucky knot: Clove hitch

Gemini

An ascendant Saturn indicates that if you are planning a sideline of your own, now is a good time to start an Ink Spots tribute band. On Wednesday, if an older person gives you advice, Jupiter is trine in your fourth house, so instead of getting your gander up (it needs its rest) just tell them to shove off and mind their own business.

Lucky ice-cream: Burberry
Lucky pants: Day-glo orange

Cancer

If it is your birthday this week, Pluto enters your fourth house on Wednesday which indicates that the celebrations may not go as smoothly as planned when a post jelly-and-ice-cream game of, 'What's the time Mister Wolf? gets somewhat out of hand leaving three members of the secretarial staff from your office with quite severe bites.

Lucky pie: Hare
Lucky Herbert: Asquith

Leo

Socially, this could be an exciting week. Wednesday may bring an invitation you'd do well to accept. Ignore any protesters that you may encounter as mice hockey is a growing sport and needs all the support it can get. By Friday, help from a quarter you least expected could give you the breakthrough in Pantomime you've been seeking for so long.

Lucky Camp: David
Lucky Rush: Eric

Virgo

Mars in your sign is likely to cause domestic upheaval this week. Expect an argument on Thursday over who gets to unload the dishwasher. Pluto transits your birthsign on Friday, so a chance meeting with Roger Moore in Argos could well see you being offered the lead role in the BBC remake of *The Woodentops*.

Lucky vole: Water
Lucky glue: Copydex

Libra

A persistent man in cricket whites will take on new significance in your life on Wednesday, whereas the New Moon in Scorpio brings only the promise of more free Eddie and the Hotrods concert tickets. On Friday you will be followed home by a swarm of locusts, so it might be worth picking up some Raid rather than just putting it down as 'one of those things'.

Lucky tribe: Crow
Lucky veneer: Maple

Scorpio

Last Monday's Full Moon stretched your patience to the utmost, so relationships of all kinds are going to suffer this week, particularly after last Friday's over fifties lunchtime special at the 'Soapy Frog' slipper-baths. On Wednesday, be ready to respond to surprising offers, particularly those involving Gruyere or Brillo pads.

Lucky dance: Lindy-Hop
Lucky cleaner: Doris

Sagittarius

Your magnetic personality has long been a legend in Rhyl, but this week it causes you more trouble than usual on Tuesday lunchtime. While passing a series of road-repair vans in Lower Regents Street, you become stuck to a transit van, and spend the remainder of the week in Porthcawl.

Lucky gesture: Token
Lucky language: Colourful

Capricorn

Just when you were beginning to despair of ever being appreciated, at last, you have been recognised in high places. This trend will continue through the week with Saturn entering your sign on Thursday when you will be recognised in three lifts, and on the escalators in Selfridges.

Lucky wire: Warrington
Lucky Penn: William

Aquarius

Your ruler, Mercury, is going to be retrograde for the next few weeks. This always makes life uncertain and difficult for you, and could lead to nagging doubts, self-recrimination, bitterness, and sleepless nights recalling past embarrassments. This is perfectly normal for anyone contemplating a new fowling piece, so try to remain calm. On Friday you may be chased from a licensed premises with a yard broom.

Lucky tree: Shoe
Lucky stock cube: Lamb

Pisces

However ambitious you are, you like to have your home to return to and then close the door. This week, however, your chest feels much better so you feel like flinging the door open and offering invitations to all and sundry, your friends, family, and even the man down the road who sells religious lawn statuary. Steer clear of Eggs Florentine on Friday, especially as you may have a hot date.

Lucky club: Army & Navy
Lucky field: Arable

Weekly Forecast for 4th to 10th July

Aries

Last month's Summer Solstice and Wednesday's Full Moon mean that you'll begin to make progress just where you hope to. Friends may be surprised when they see the radical changes you make to both your attitude and your choice of footwear. After all, it's not everyone that has the calves for platforms, and it's time to flaunt them.

Lucky posture: Improved
Lucky preservative: Alcohol

Taurus

Great changes are afoot for you. This week the Sun transits Pluto and is followed immediately by a Full Moon in Capricorn. This indicates that you will have a unique opportunity to find your true self. It might be best to start looking in the cupboard under the stairs, where you already have a lot to hide.

Lucky horn: Cor Anglais
Lucky plum: Victoria

Gemini

This week, the Full Moon following the Summer Solstice means that a strong woman with a plate of eclairs may feature in your chart around the middle of the week. Although some uncertainty is indicated by the presence of Neptune in your fourth house you will at least have some idea where you stand. If you don't, try asking a grown-up.

Lucky Penn: Sean
Lucky stock: Night scented

Cancer

You're likely to enthusiastically greet all the unexpected opportunities that are heading your way this week. Keep your wits about you though, and your feet on the ground – or at least off of the desk, as Pluto transiting your seventh house indicates that on Thursday when examining one of the drawers for a mislaid niece, you discover a packet of steroids and an 'elephant-ears' thong in a secret drawer which can only mean that it is a rare 18th century Chippendale.

Lucky choice: Sophie's
Lucky plunge: Icy

Leo

After an exhausting week off, your ruler, the Sun, moves into dynamic Taurus, filling you with unusual impulses. You may make a new friend this week, or find that an old obsession becomes interesting again. You could meet someone from far away, or even turn to a life of contemplation in a monastery. Important days will be Wednesday and Thursday, when you see a side of someone you don't like, or possibly just a person you don't like from one side.

Lucky look: Baleful
Lucky tone: Mellow

Virgo

As Thursday's trine between Venus and Saturn approaches, a close friend may bring you good news about those split-ends. With Venus rising, there's a good chance of new support trousers on Saturday, so make sure you are in the right frame of mind for a fitting. Watch out for a drama involving blue cake on Thursday.

Lucky reaction: Surprise
Lucky event: Three Day

LIBRA

At long last things for Libra are looking up. On Tuesday, a lovely aspect between the Sun in Virgo, related by the common rulership of the Moon in the area of your chart affecting foot-care means that you will finally discover an alternative for the flick-knife and the cheese-grater that you normally use to keep your hard-skin under control.

Lucky flooring: Rush mats
Lucky Dance: Charles

SCORPIO

With Mercury high in the section of your chart that governs your career you should buy the kaftan you've been admiring for weeks – a striking outfit will come in handy when you need to start looking for another parish. On Thursday you will run into a former colleague. Fortunately you both have comprehensive insurance and sympathetic brokers.

Lucky topping: Onion rings
Lucky roofing: Norfolk reed

SAGITTARIUS

Friday's New Moon in your sign will have helped things along a little, and this should be a friendly and sociable week, but don't force your views about shoe-polish on others now, as they may not be receptive. The weekend could see someone mysterious come into your social circle, possibly someone who doesn't even care for leapfrog.

Lucky sprinkler: Rotary
Lucky snack: Dubbin

Capricorn

Relationships with friends and neighbours will go more smoothly, (particularly with the lady at number 32 whose husband works nights) in the weeks to come as Mars transits your fourth house. There may be some confusion to do with your work, and it is possible that someone in your vicinity is not being strictly open and upfront over a possible move to sell all company staff into white slavery.

Lucky clouds: Cirrus
Lucky stitch: Running

Aquarius

Your impulsive and emotional nature may get the better of you at the start of this week. With mischievous Neptune rising, and Mars in your sixth house, it is possible that someone may discover the pillowcase you keep in the bottom drawer of the filing cabinet, stuffed with woodlice. On Thursday you may well discover a new enthusiasm for being stand-offish after 3pm.

Lucky light: Ultra-Violet
Lucky trousers: Lycra

Pisces

Your winning way with a variety of everyday objects is now beginning to take on legendary status, but on Wednesday a trine Pluto indicates that you are on the verge of an important discovery, and by the end of the week, using only two cocoa tins, a length of hairy string, and a modelling knife, you will have constructed a powerful telescope with which it is possible from your house to see the nurses accommodation blocks at the local hospital.

Lucky excuse: Astronomy
Lucky alibi: Research

Weekly Forecast for 11th to 17th July

Aries

Recently you've had to change your plans, when doors that should have opened remained firmly closed. With Jupiter entering the section of your chart governing warranties and guarantees, it may be a good time to take back the crowbars and bolt-croppers you have been using. Your choice of a monogrammed marigold at the scene of each robbery may have consequences.

Lucky pick: The Locksmith's friend
Lucky jersey: Striped

Taurus

Surprising or exciting news comes to you this week. While it may seem as if you've been thrown into a spin, that may just be the first sherry of the day. On Wednesday, comforting Venus enters your birthsign bringing with it all sorts of warm and rounded experiences. You should seize these with both hands as

Neptune transits your birthsign at the end of the week, which indicates the unwelcome return of sporadic belly-button failure.

Lucky slippers: Ruby
Lucky rabbits: Cottontail & Peter

Gemini

A lovely trine between the Sun in Virgo and Jupiter in the area of your chart affecting body lotion indicates that during a visit to Broadstairs on Tuesday, while assisting a distressed research assistant with one of the reasons for her distress, you become entangled in fishing nets and have to spend the remainder of the week on the dogger bank, where you are surprised to discover several parked cars with steamed-up windows in a woodland clearing.

Lucky Glo: Hunniford
Lucky varnish: Matt

Cancer

An interesting aspect between Saturn and Pluto this week could mean that you are contemplating an image change. Having been a slave to fashion all your life, it will come as something of a surprise to your colleagues to see you giving away your fine collection of Italian silk ties to charity. On Wednesday, a rough and ready little man with abundant dandruff will offer to Hoover your turn-ups for a pound.

Lucky hand: Royal Flush
Lucky weakness: Black-Jacks

Leo

Neptune rising on Thursday means that the papers may get hold of the story about The Shadow Defence Secretary, the slow cooker and a gallon of maggots. Pass it off as lightly as you can, the waning moon indicates that the Public Record Office is unlikely to press charges this time. Friday is a good day to buy a replica covered wagon.

Lucky cheese: Chevre Blanc
Lucky fencing: Waney-lap

Virgo

An interesting week, made all the more exciting by a wonderful trine between Mercury and your ruler, the Sun. This indicates that the long wait for your dream film role is finally over. The opening scene of the next James Bond epic will see those distinctive features on the big screen at last as you utter the immortal words, 'Evenin' Standard Guv?' to the master spy as he hurries past.

Lucky lifting equipment: Block & Tackle
Lucky accessory: Bicycle clips

Libra

Your ruler turns retrograde on the same day that the Sun moves into Libra, which is just typical with visitors due, that nasty black-treacle stain on the hall carpet and the tumble-drier full of goose-down. A minor fungal infection may not be quite such a bad thing as you are able to sell the fruiting bodies to a French farmer's market at a very decent price per kilo.

Lucky blemish: Warts
Lucky ringtone: *Bullseye* theme tune

Scorpio

On Thursday, a rare and wonderful trine between Mars and Neptune is perfectly placed to enable you to sleep right through the alarm and turn up for work looking like one of those men you see in the park talking to brown paper bags. Venus joining forces with Jupiter this week indicates that you will at last fulfil your dream to become a Danish language mime-artist and exchange your mundane life for one filled with pickled herrings and pleasure.

Lucky haircut: Mohican
Lucky weakness: Salt tablets

Sagittarius

It has been a demanding and difficult summer for many Sagittarians, and you've had to endure one blow after another for some weeks, but on Wednesday Mars is poised to move into the practical Capricorn, which should provide some respite. On Friday, a man with a bushy beard will try to persuade you to take out cat insurance – don't be taken in – you don't have a cat.

Lucky angle: Acute
Lucky fish: Whiting

Capricorn

You have always liked being the centre of attention, and the intervention of Chiron in your sign means that your talent for dancing will at last be recognised and you will be invited to host the 'White Heather Club' this New Year's eve in Scotland, opening the show by dancing a spirited 'McPherson Strut' with Moira Stewart.

Lucky sensation: Dizziness
Lucky vegetable: Neeps

Aquarius

You've not been sleeping as well as you might of late, and it's no wonder. A 'Sou-wester' does not make for comfortable nightwear. However, as boisterous Saturn transits your seventh house on Tuesday this is all set to change. While shopping for new restraints, you will spot the very thing to improve those restless nights – a dribble-proof pillow.

Lucky collar: Astrakhan
Lucky sandwich: Ross-on-Wye

Pisces

This week, with Pluto entering your birthsign you will see the start of an entirely new phase in your life when on Monday, following a very good lunch, you set your mind on becoming a Dervish. This however proves a rash decision as you both figuratively and literally take a turn for the worse while trying to counteract severe room-spin.

Lucky Cup: C
Lucky gland: Adrenal

Weekly Forecast for 18th July to 24th July

Aries

Even with Mercury in retrograde, the signs are very good indeed for any Gemini whose work involves attending a lot of meetings at which you nod sagely and say 'I'll get back to you on that'. Saturn rising on Wednesday indicates that jealousy and power struggles will loom large while visiting the cold-meat counter in Morrisons.

Lucky paper: Wall
Lucky hole: Wookey

Taurus

A trine Venus means that this could be rather a low spot in your love life. The only way to get through this difficult week is to be philosophical and whatever you do, don't take things personally – even the comment about hummingbirds. On Friday, Pluto in Aries means that you will receive a mysterious package containing a book of pressed starfish. Try to remain stoic.

Lucky accent: Nigerian
Lucky affliction: Athlete's foot

Gemini

This is a very good week for making your home into a wonderful haven for wildlife. Unpredictable Pluto conjoins with the Moon in your fourth house to give you the irresistible urge to build shelters for wild creatures under your eves. As you don't have eves, you may be tempted to keep them about your person. This could lead to difficulties at airports.

Lucky hair-gel: Golden shred
Lucky soil: Medium loam

Cancer

This is a great week for communication for Cancerians. Whether you're talking things through at work, making business plans or simply trying on new latex evening-wear, you should make sure that others appreciate your important points. On Wednesday a square Pluto means that you will run out of gob-stoppers mid-morning and have to make do with Love-Hearts.

Lucky fly: Tsetse
Lucky envelope: DL Window

Leo

This week you'll find that other people have unexpectedly come round to your way of thinking, and you can make changes to everyday matters that will radically alter the accepted way of doing things. The Moon in Libra means that you may struggle at first to get people to accept an hour of community whistling from 8am, but you should persevere.

Lucky snack: Elvers
Lucky belt: Fan

Virgo

At times this week it may feel as though you are being required to juggle several balls at once. This may have more than a little to do with the cut of the new staff overalls. Saturn rising midweek may leave you feeling as if you are in a state of limbo, but you will find that this is due to you having done the buttons up incorrectly.

Lucky leg: Quarterfinals
Lucky hinge: Queen Mary

Libra

With the Sun, your ruler, transiting mysterious Neptune it's no wonder that you are wracked with self-doubt, however, on Wednesday, Mercury planet of communication enters your third house bringing with it the irresistible urge to go back to the long held dream of writing the musical comedy version of *Angela's Ashes*.

Lucky lining: Silver
Lucky phrase: Stay back! I got a shooter

Scorpio

With so many planets crowding into your sign you will be feeling full of confidence and ready to take on the world. This may come in useful later in the week as on Thursday you may well have a spot of bother after you are caught sticking up for a lady colleague at the company sauna night.

Lucky flooring: Linoleum
Lucky pattern: Interference

Sagittarius

Unpredictable Aquarius is about to conjoin with Neptune. As you are prone to excess nasal hair, the Full Moon in challenging Saturn on Wednesday means that your clippers may finally give up the ghost. This will come as a bitter blow to your partner who was particularly looking forward to topiary practice at the weekend.

Lucky instrument: Flugelhorn
Lucky cartoon: The Jetsons

Capricorn

Once again, troublesome Mars means that you won't get all of your ironing done for the weekend. At best you'll manage a few shirts. To make matters worse, you will run out of spray-starch after the first one. On Friday the Sun challenges your ruler placing the interesting asteroid known as Chiron in a celestial position known as the dangling wampum.

Lucky shrub: Variegated Dogwood
Lucky snack: Eggs Benedict

Aquarius

Toward the end of this week Venus transits your birthsign

bringing with it more of the good fortune that has so far this month led to you winning seven pounds fifty at five-card brag, and your own weight in shaving foam in a newspaper crossword competition. On Wednesday that pain in your knee might give you a bit of gyp during salary negotiations. Try sitting rather than kneeling.

Lucky gesture: Gallic shrug
Lucky whim: Barley-sugar

Pisces

This week a trine Venus on Wednesday means that this could be rather a low spot in your already troubled love life. The only way to get through this difficult week is to be philosophical and whatever you do, don't throw out the equipment, or have the soundproofing or deadlocks removed just yet.

Lucky manner: Offensive
Lucky belt: Van Allen

Weekly Forecast for 25th to 31st July

Aries

An interesting week, made all the more exciting by a wonderful trine between Mercury and Uranus – planet of surprises. Therefore on Tuesday it will not come as a shock that you will meet a charming attractive lady who will give you both her phone number and a particularly stubborn fungal infection.

Lucky spasm: Nervous tic
Lucky affectation: A monacle

Taurus

Your ruler, the Sun, conjoins with Mercury planet of communication on Monday, so it looks like a particularly busy week ahead. A slim woman with twinkling feet and a vacant smile may make your life a misery toward the end of the week. On a brighter note, a group of Television Evangelists will approach you with money and several potential Bridge partners.

Lucky starter: Soup of the day
Lucky herb: Miller

Gemini

You have long been an admirer of the more aesthetically pleasing things in this world, but the malign influence of mischievous Pluto combined with an unguarded comment about Fiona Bruce's legs may have led to disagreements. However, you can look forward to some very positive developments in your chart once the plaster is off and the stitches are out.

Lucky ointment: Arnica
Lucky bandage: Crepe

Cancer

An unusual start to the week is indicated by a square Jupiter entering your birthsign. This probably means that the lucrative work you have been receiving as a body stand-in for some specialist films may be coming to an end as despite the elaborate make-up, people are starting to suspect.

Lucky sheet: Balance
Lucky excuse: Damp kindling

Leo

As the Sun, your ruler, moves into the influence of dynamic

Mercury you will be filled with the kind of inspiration you've not experienced for years. Time to start planning a holiday. On Thursday a woman of substance will tread on your corns while dancing the Lindy-Hop.

Lucky oxide: Copper
Lucky line: Waterloo

VIRGO

Tuesday or Wednesday could see you invited to an entertaining social event by a close friend at which you shine. Take extra care to ensure that the dishwasher marks are completely removed from the champagne flutes before serving, and don't eat too much pastry while circulating with the petit-fours as you know what it does to you.

Lucky pâté: Coarse Brussels
Lucky particle: Neutrino

LIBRA

There has been a gathering of planets in your sign in recent weeks bringing with them a series of peculiar events. These have mostly involved marmite soldiers being pushed through your letterbox, anonymous peanut butter sandwiches sent to you by recorded mail, and phone messages suggesting you eat more fibre. On Thursday all will be revealed when the culprits turn out to be nothing more than a particularly zealous group of Hovis Witnesses.

Lucky bat: Cricket
Lucky Ball: Michael

SCORPIO

Although it's something you've probably not noticed over the years, you remain nearly as popular with the ladies as Michael Fish. Mars entering your fourth house on Wednesday can only help to enhance this reputation. However, a misunderstanding during an interview in which a new position at the tennis-club is being discussed, does result in one over-enthusiastic lady applicant being used to transport grass cuttings to the compost heap.

Lucky smell: Marzipan
Lucky knot: Gordian

Sagittarius

Stubborn Pluto entering your sign on Thursday means that your plan for an all-action accounting holiday in Montana may involve rather more action than you first anticipated. On speaking to a representative from the company on Tuesday, you discover that there are significant differences between rounding up cattle, and rounding up numbers on a spreadsheet.

Lucky flower: Dahlia
Lucky rub: Fiery Jack

Capricorn

Too much rich food, combined with a particularly fearsome sprout vindaloo on Saturday night left its mark on your digestive system. This manifested itself early on Sunday morning in the most spectacular way yet. Fortunately for you, most of the structural damage was attributed to a small fire at a local chip shop. Saturn enters your birthsign on Wednesday and leaves a nasty mark on the kitchen lino.

Lucky penguin: Chinstrap
Lucky soup: Cream of Magnesia

Aquarius

Over the next few days you may find, mainly with your shins, that your furniture will be rearranged. This will be done in accordance with the principles of 'feng shui', an ancient Chinese philosophy whose name means, literally, 'cough-up round-eye'. This teaches us that where we locate our household items affects our happiness by controlling the flow of 'chi', which is a life force that is always around us, everywhere, all the time, like Ant & Dec.

Lucky instrument: Blunt
Lucky Cartwright: Hoss

Pisces

Money is a little tight this week, and with the Full Moon in challenging Saturn at the end of the week, even though you've set your heart on it, it looks like you'll have to wait another month for the new awning for the caravanette. Console yourself with the good news that bottled gas is actually cheaper than it was this time last year.

Lucky link: Missing
Lucky Circus: Oxford

Weekly Forecast for 1st to 7th August

Aries

The effect of Chiron transiting your fourth house means that on Wednesday you will have a very strange dream in which you find yourself being lowered into a cauldron filled with goldfish by Jeremy Clarkson. This clearly indicates that you should consider giving up anchovy and radish sandwiches for supper. Community folk dancing may prove crucial at the end of the week.

Lucky tart: Pear and almond
Lucky almond: Marc

Taurus

Communication may come on an unspoken level, given today's conjunction between Venus and the Moon. People will be saying a great deal with their eyes, and a great deal more with their letters to the authorities and the occasional wrapped brick. You can safely ignore the taunts, as you know full well that it's only water-retention.

Lucky Womble: Tobermory
Lucky alias: Edith Evans

Gemini

Gemini is, as everyone knows, a sign of duality. This can mean that you have two ways of looking at everything, or that nice things just happen to come along in pairs. On Thursday, Pluto, planet of surprises enters your birthsign which means that the new girl in the staff canteen may present you with a pair of large cappuccinos instead of the espressos you are used to. Accept them with good grace.

Lucky language: Algonquin
Lucky priest: Judas

Cancer

Now that Mercury is once more direct, the pace of life quickens. With your zest for life, you are in your element, which curiously enough is Iodine. Later this week, you are likely to once more indulge your passion for bath games, but are likely to find that combining a hot bath with a thousand-piece jigsaw puzzle proves impractical, and leads to a reverse print of Willy Lott's Cottage on your legs and a blocked plughole.

Lucky pasture: Verdant
Lucky club: Knobkerrie

Leo

The Sun, your ruler, enters erratic Neptune this week bringing with it the possibility of an exciting new venture, or possibly an exciting old venture with a new twist. Your mood toward the end of the week may be surprisingly buoyant as Venus is in fabulous aspect with benevolent Jupiter. However, it may just be down to the new slippers you got in the sales.

Lucky fish: Rock-Eel
Lucky plaster: Breathable

Virgo

Unpredictable Aquarius is about to conjoin with Neptune so you may feel more than a little paranoid this week. The best advice is to always check your sock drawer for venomous snakes before bed rather than in the morning when you are less alert. On Thursday a career change is indicated by Saturn rising. This means that you can expect to be fired by the weekend at the very latest.

Lucky affliction: Postman's knock
Lucky fleece: Golden

Libra

Pay attention to the details as you establish a new routine for

August and life is likely to go more smoothly than of late. The only reason that you got away with no trousers on Friday was that it was dress-down day. Mercury in opposition to Saturn means that Wednesday is a great day for doing things with other people, so there's just time to dust off the equipment and book the ticket to Ipswich.

Lucky crisps: Spit 'n' Sawdust
Lucky tiles: Polystyrene

Scorpio

Cheese-on-toast was a mistake for breakfast this morning, and will come back to haunt you later in the day. Mars and Saturn conjoin in your fourth house on Wednesday and, as usual, leave bits all over the hall carpet. A cross-looking woman misusing a trampoline is indicated by a trine Neptune toward the end of the week so try extra hard to stay out of trouble.

Lucky cake: Battenberg
Lucky cat: Mackerel Tabby

Sagittarius

It is becoming increasingly clear that the relentless pressure of work is starting to take its toll on your normally ebullient nature. You should give serious consideration to having some time off – however much it goes against your nature. On Friday, Mars rising may mean that your favourite cash-point is struck by a meteorite.

Lucky element: Iodine
Lucky affliction: Swamp ague

Capricorn

A difficult start to the week for anyone with the Sun strong in their chart. It's a particularly arduous afternoon on Wednesday when your work on the latest sales figures from Belarus is disrupted by unruly drunks. Try to pass it off with your usual aplomb (it's the one with the brass handle).

Lucky mash: Celeriac
Lucky grip: Masonic

Aquarius

An interesting aspect between Saturn and Pluto this week could mean that you are contemplating starting a campaign to return Blakeys to their rightful place under the heel of the working man. On Wednesday, a Native American man may offer you a suspiciously low-priced Hillman Avenger. Despite your well-known enthusiasm for the model don't be tempted.

Lucky elbow: Tennis
Lucky view: Jaundiced

Pisces

Neptune rising on Thursday means that you may at long last expect some sort of reward for the long years of service dedicated to the welfare of your fellow human. Even though you are best known as 'that one who smells of nutmeg', you will receive a Postal Order for £5, a photocopied certificate, and a commemorative mug declaring that you have been officially recognised as a 'boon to the elderly'. Try not to let it go to your head.

Lucky cheese: Port Salut
Lucky fencing: Picket

Weekly Forecast for 8th to 14th August

Aries

An interesting week, made all the more exciting by a wonderful trine between Mercury and Uranus – planet of surprises. On Monday you are approached by a man posing as an Arab Sheikh and attempting to lure you into an elaborate scam to sell cut-price inflatable war-memorials to the Canadian Government. On Friday, toasted teacakes make a welcome return to the staff canteen.

Lucky tip: 15%
Lucky mask: Seaweed

Taurus

Mercury, planet of communication, moves retrograde on Monday so this is not a week to rush into things. That isn't so easy for Gemini, for no sooner have you got an idea than you are dashing to put it into action. Bide your time, particularly when it comes to both box kites and new relationships. Keep personal items to yourself until you feel the time is right – perhaps over lunch – to lay them on the table.

Lucky fittings: Gold-plated
Lucky spanner: Adjustable

Gemini

This week may start with events that seem discouraging, but in fact time is on your side. Whatever setbacks you experience in the short term, benign Jupiter is all set to make sure that any charges that are pressed are dismissed before the case ever comes to court. On Thursday you are surprised to be elected 'Pearly King' of the Balls Pond Road but following your enthronement, will be carried off by Magpies.

Lucky grape: Zinfandel
Lucky aroma: Marzipan

Cancer

Mars moves into your birthsign on Tuesday meaning that this week you'll have energy and imagination and should be able to put both to excellent use when thinking up new places to hide from the woman from number 14, who has the capacity to drain the life force from you faster than daytime television.

Lucky soft drink: Cresta
Lucky disguise: Rev. Canaan Banana

Leo

Jupiter transits mysterious Neptune this week bringing with it a slightly introspective mood. You may find yourself wondering whether you'll ever be anything more than a man well known in the Hillingdon area for his first-class imitation of an angry swan. A man of the cloth may prove troublesome over lunch on Monday.

Lucky surface: Enamel
Lucky pie: Rabbit

Virgo

Everything goes well this week until Monday morning when at noon you come under the influence of Mercury, which – as usual, means you'll get into another protracted argument about gravity. On Friday you may feel a sudden urge to stop and pick wild flowers by the side of the road but as it's the

M4, this proves easy to resist.

Lucky Flann: O'Brien
Lucky tart: Custard

LIBRA

With Mercury in troublesome Pluto, this week is not a good time to travel. Even though you took the precaution of getting the Morris Marina serviced, you can still expect a torrid time before you finally get to Gillingham. On Thursday you may feel a little under the weather after seeing someone wearing a black suit with brown shoes.

Lucky Marten: Pine
Lucky Pine: Courtney

SCORPIO

Wednesday's aspect between Venus and Neptune will probably mean that you will have a misunderstanding involving tile adhesive at the Department for Work and Pensions. A man with a limp handshake and brogues may attempt to interest you in a set of shop-soiled kitchen scales.

Lucky colour: Magenta
Lucky alarm: Smoke

SAGITTARIUS

Your natural modesty sometimes does you no favours. Though you have been out of the office for the last few weeks, only your closest friends are aware of your triumph in Australia. Even the return of the coveted Patterson Challenge Shield to the company trophy cabinet fails to attract comment. Console yourself with the thought that only a select elite is ever called on to compete in the demanding Jacobs Creek 50 litre freestyle.

Lucky fairy: Puck
Lucky composer: Orff

Capricorn

On Monday, Mercury rushes through the area of your chart concerned with empowerment and authority. Normally you are not very far up the chain of command when it comes to making decisions and often catch yourself sneaking a ginger biscuit out of the packet and eating it furtively in your shed. This week all that is about to change. You run out of ginger biscuits and have to make do with a fig roll.

Lucky costume: Doublet & Hose
Lucky stroke: Forward Defensive

Aquarius

On Tuesday a rare and wonderful trine between Saturn and Neptune is perfectly placed to enable you to make the most out of your newly discovered ability to make a noise like a wounded muskox. The coming weeks should see the demand for your voice-overs leap like a startled Brownie. Toward the end of the week an accident with a sausage sandwich and a new shirt is indicated.

Lucky occasion: Auspicious
Lucky atmosphere: Strained

Pisces

Your recent enthusiasm for Cornish pasties has left you carrying a couple of pounds more than is good for you. But it's not until you discover that you have been nicknamed 'two mirrors' that you finally decide to do something about it. A new 'perfect body' exercise regime will really have you jumping through hoops. On Thursday you will discover that this is due to a mix up when enrolling between Chippendales and Chipperfields. Make sure you've flossed before allowing the instructor to put his head in your mouth.

Lucky gazelle: Thompsons
Lucky voltage: 110v DC

Weekly Forecast for 15th to 21st August

Aries

Having been a slave to fashion for the last few years, some doubts have begun to creep in about whether you can still cut the mustard. However, a therapeutic visit to Lakeside restores any doubts you may have had about your good taste. Resist the urge, however tempting, to acquire a 'his 'n' hers' sun-strip for the car.

Lucky floor: Parquet
Lucky Wall: Max

Taurus

An adventurous week starts on Tuesday when Mars enters your fourth house and a mysterious letter in a spidery hand reveals the dreadful family secret you have hidden for so long. Fortunately, the responsibility exemption certificate you received last Christmas covers both this and Thursday's summons for impersonating a Coastguard in a built-up area.

Lucky stuffing: Horsehair
Lucky flatfish: Skate

Gemini

Venus transiting your ruler the Sun this week indicates that a strong creative urge manifests itself on Wednesday. While toying idly with your socket-set you come up with thc notion of a set of handicraft tools guaranteed not to work any better when someone impatiently snatches them away after watching your feeble efforts at DIY.

Lucky Stand: Custer's Last
Lucky past: Chequered

Cancer

Your lifelong dream of an online matchmaking service for the over sixties takes a bit of a knock on Tuesday, as mischievous Pluto transits your birthsign. The advertising agency doesn't feel that the name 'Carbon Dating' conveys the right message for your potential clientele. Be extra careful on Wednesday if you have to drive, operate complex machinery or gargle with rodents as Mercury turns retrograde just after lunch.

Lucky fencing: Chain-link
Lucky county: Corsetshire

Leo

A difficult trine between Pluto and Mercury, planet of communications could mean a confusing week ahead. When visiting your Doctor for a routine check-up on Thursday, you generously decide to complete an organ donor card. However, it is worth pointing out that the reason that the spaces on the form are so short is that the transplant authorities are anticipating words like 'Kidney' and 'Liver' rather than 'Bontempi'.

Lucky hinge: Evadne
Lucky flange: Swept Elbow

Virgo

A square Neptune on Wednesday sees you cornered in the lift by a garrulous colleague keen to relate a tale of heroic, yet calm and collected confrontation, of the taxman. As this will be the seventeenth time you've heard the story you could find yourself hitting the duty-free Retsina before lunchtime.

Lucky bathroom suite: Taupe
Lucky table: Drop leaf

Libra

Unlike those with the fixed signs strong in their charts you may find Neptune troublesome early in the week. On Tuesday your pre-occupation with the 1960s *Stingray* television programme leads you into conflict with the authorities once again. It's true that your 'aquaphibian' impression is second to none, but filling the cavities in your double-glazing with water and tropical fish, and lounging in the window dressed as 'Marina', gets you into trouble with both Greenpeace and the RSPCA, so try to tone it down.

Lucky songbird: Spotted Pie-catcher
Lucky affliction: Guppies

Scorpio

You've always felt, like Socrates, that the unexamined life is not worth living. However, borrowing a high-powered broadcast-quality video camera from work while the lady over the road has removed her net-curtains may bring the philosophical nature of your motives into question this week. Mercury, planet of communications, enters your sign on Wednesday, so if you need a decent excuse, wait until then.

Lucky lift: Clean and jerk
Lucky pastry: Filo

SAGITTARIUS

Money is a little tight this week. With the full moon in challenging Saturn and the question being asked about when you are going to get a nice little earner on the side like that chap with the wonky eye, it looks like you'll have to wait another month for those tyres for your caravan. This will come as a blow to your partner, who was particularly looking forward to Lowestoft next week. A yodelling accident is a distinct possibility on Thursday.

Lucky adjective: Crispy
Lucky outlook: Fey

CAPRICORN

The combination of the Moon entering your fourth house and the rising temperatures means that you may be feeling a slight restlessness at work; not so much that you want to run away to sea and spend the rest of your life under an assumed name working as a crewman on a cockle-dredger, but it may be worth investing in new oilskins just in case.

Lucky fork: South
Lucky pastry: Rough-puff

AQUARIUS

An excellent week during which you will be presented with innumerable opportunities to laugh at others' misfortunes, poke fun at those with too much home-contents insurance and draw spectacles on pictures of leading Academics. On Thursday, a stroke of luck in the canteen raffle will see you receive a jar of hand-lotion and some high-altitude climbing equipment.

Lucky approach: Cautious
Lucky glass: Stained

Pisces

Does it sometimes feel as if you're doing all the work and others just sit back and let you get on with it? The saying 'If you want a job done properly, do it yourself' might well have been said first by you. The rest of the team just don't seem to be pulling their weight. On Friday a bony woman with an intense stare takes exception to your second-best dancing trousers in the local wool-shop and has you barred.

Lucky collar: Eton
Lucky implement: Laundry tongs

Weekly Forecast for 22nd to 28th August

Aries

Like most Aries, you live in the fast lane, working and playing very hard. However, you must remember to build in some rest and recovery time as it's likely that if you carry on at the rate you've been maintaining for the last week, you won't have any energy left for Pro-Celebrity Banger Racing on Thursday. A chance discovery of a lemon bon-bon in a jacket pocket on Friday finishes the week on a high.

Lucky beans: Borlotti
Lucky clam: Giant Pacific

Taurus

With the last of the August sales now with us and the Moon in Leo, there is an almost frantic energy about your efforts to secure those last-minute bargains. On Wednesday your enthusiasm for a leopard-print bikini spills over and you let yourself down by becoming involved in an unseemly scuffle with the Duchess of Argyll at the clearance rail in Peacocks. On Friday a casual remark to a local ombudsmen may lead to an unexpected offer of shenanigans. You should reject this, as you know what Irish food does to you.

Lucky hobby: Fluffer
Lucky bird: Sword-finch

Gemini

The intervention of Chiron in your sign on Tuesday can only mean that after all these years your Coconut Matt Monro doormat finally catches the public's attention. It is not only hard-wearing, but welcomes you home with some of the great crooner's all-time classics. If you fail to wipe your feet it makes the sound of galloping horses. A snip at £19.99 and a joy to hoover.

Lucky disguise: Orla Guerin
Lucky punishment: Nutmeg

Cancer

The Moon joins Venus, the planet of relationships, in the area of your chart that governs corduroy trousers on Wednesday. It looks likely that you will find yourself in a confined space, possibly a lift or maybe a packing crate, with a similarly dressed but highly-attractive potential partner. Your eyes will lock, as indeed will both pairs of corduroys, so you may find yourself tangoing to the closest fire station before you can be separated.

Lucky mood: Kittenish
Lucky game: Kerplunk

Leo

Jupiter has moved into a very creative and vital part of your chart this week. If you set your mind to something, you can do it – regardless of what those court orders say. What's more, your ruler, the Sun, is close to Mercury at the minute so a sensible hat and factor 20 would be a worthwhile investment. On Thursday lunchtime you may be signed up as a Jack Hargreaves tribute act.

Lucky condiment: Mint sauce
Lucky ranch: The Ponderosa

Virgo

The Sun arrives in your sign this week. If it's your birthday next week, it will bring wellbeing and good fortune, and a benign influence that will permeate the whole of the next twelve months. If, however, your birthday falls this week, the best you can expect is for a Cornish-looking gentleman to show you a suspicious rash on the bus on Thursday lunchtime.

Lucky device: Phillips Babyshave
Lucky lawman: Snoop Deputy Dawg

Libra

The Sun is trine Saturn this week, which indicates that you should make sure that you have both an up-to-date map of Slough and a three tier cake-stand about your person on Tuesday lunchtime as they will help to avert a dangerous PowerPoint overdose at an important board meeting.

Lucky snack: Sardines on toast
Lucky adjective: Wonky

Scorpio

Neptune transits mysterious Pluto later this week, bringing a very real sense of urgency to your sponsored slim. Your target weight-loss of 20 kilos by Christmas is starting to look a little ambitious as you've not passed up many Wagon Wheels of late and your devotion to the mid-morning Hobnob is almost legendary.

Lucky legs: Cabriole
Lucky lunch: Sudden fried chicken

Sagittarius

A shameful episode while shopping at Bluewater on Monday means a quiet week during which you would do well to ensure that you have finished your tax return. The New Moon in Aquarius indicates that new hall curtains are on the cards. Make sure your dry-cleaning is up-to-date, as the shop will be burned down by goblins on Wednesday afternoon.

Lucky snack: Sugar-mice
Lucky trousers: Leather

Capricorn

You may have become disillusioned with other people in recent weeks and felt like focussing on the wonders of the natural world. Your recent investment in a butterfly net proves to have been worthwhile when, while hiding in the undergrowth as usual on Tuesday, you catch an example of a completely new species, the Essex yellow. Easily identified by its markings, which look like a tattoo of a fat girl on each wing.

Lucky teeth: Premolars
Lucky mustard: Dijon

Aquarius

The early part of the week may bring you an unexpected message from the past – perhaps an old acquaintance, or possibly a friendly bus inspector. Take extra care on Friday as a square Pluto indicates that a small woman with a fearsome grip enters your life and will try to persuade you to take up either lawn-bowls or a life of piracy.

Lucky cloud: Nimbus
Lucky drink: Warm water

Pisces

It's a busy time. You may start the week in a whirl of activity, but on the whole it is positive and productive until Tuesday lunchtime, when you fall into bad company in the local Waitrose and wake up buried up to your neck in the central reservation of the North Circular at Wembley. On a positive note, you make a small fortune from slowly passing drivers as an unusual busker.

Lucky colour: Prussian blue
Lucky vitamin: D-Minor

Weekly Forecast for 29th August to 4th September

Aries

Mars entering changeable Libra indicates that your recent business venture 'The Butch-Cassidy mobile hairdressing salon' is proving not to be a hit with potential clients when they discover that sitting on your handlebars while having their hair trimmed and negotiating the West-End traffic is a far from relaxing experience. A long-forgotten loved one will make an unexpected appearance on Wednesday. Buy the negatives at any price.

Lucky joint: The hip
Lucky draw: Western

Taurus

Thursday's trine Saturn indicates that your plans for setting up am all-terrain wheelbarrow racing course in Slough are likely to be rejected by the local authority on the basis that the activity is unlikely to generate enough noise and will therefore not fit in with other amenities in the area. On Friday you will discover that you are allergic to Daventry.

Lucky lotion: Calamine
Lucky tea: Camomile

Gemini

It is an undeniable fact that for years you have been nearly as popular with the ladies as Ken Bruce, but time is starting to take its toll. Mars transiting Aquarius on Wednesday combined with your unhappiness with the recent centrefold pictures in the *Catholic Herald* will give you the impetus to seriously look at a strict regime of exercise and skincare for at least ten minutes, before reverting to your normal sloth, kebabs and strong drink.

Lucky stretch: 18 months
Lucky peas: Marrowfat

Cancer

A square Mars in Uranus may mean an uncomfortable start to the week. On Wednesday Pluto entering your fourth house indicates that your position on the company's cultural integration committee may be at stake, as the suggestion that the seasonal celebrations might include a 'wet burka' competition', causes one or two raised eyebrows at board level. On Friday, a brown cardboard box is prominent.

Lucky flares: Distress
Lucky platforms: 3 and 11

Leo

Thursday's conjunction between your ruler the Sun, and dynamic Mercury means that there is a strong chance you'll need to steer well clear of fussy-eaters or anyone with dense nose-hair. Toward midweek you will be less troubled by a persistent tapping noise from under your desk than in recent months, although hard skin will still prove problematic. Don't be tempted by a bargain steamroller; it may not be all it seems.

Lucky brand: Coca-Cola
Lucky tiles: Polystyrene

Virgo

A difficult week made all the more unpleasant by the presence of troublesome Pluto. A chance meeting with a city-based head-hunter on Tuesday will see you making a late bid for a position as ceremonial goat-stretcher to the Windward Isles. An ill-starred cheese and onion pie may return to haunt you on Friday night.

Lucky owl: Brown
Lucky cake: Kendall-mint

Libra

A note of caution this week. Retrograde Saturn means that if you deal with others individually, success will almost certainly greet you. Deal with them as a group and you will be attacked and eaten by wild dogs in World of Leather. On Thursday, a distant relative will send you a first-class recipe for 'mutton surprise'.

Lucky adverb: Meanwhile
Lucky spoon: Ladle

SCORPIO

The New Moon in Venus indicates that your continued enthusiasm for late nights and willowy women is doing you no favours. This is particularly evident on Tuesday morning when you start to get out of breath after just three games of squash. On Friday, a desperate struggle over the last Eccles cake will see you escape with just light flesh wounds.

Lucky dynasty: Ching
Lucky herb: Catnip

SAGITTARIUS

The arrival of mysterious Neptune in your birthsign midweek indicates that others may begin to appreciate your creative skills. On Tuesday, a trine Saturn means that you will be consulted over the name of a new winter warming hot breakfast cereal. Although your suggestion meets with initial approval, the manufacturers feel that 'Captain Oates' does not convey the message they'd hoped for.

Lucky Green: Phillip
Lucky hole: Wookey

CAPRICORN

A pleasant start to the week as benevolent Jupiter enters your seventh house on Monday. On Tuesday your bus driver decides to do something to cheer up that wearisome trip into work when he surprises the passengers with a magnetic billiards set. This combined with the stick-on Lord Coe for the bus window with 'moving eyes and real fur', really sets you up for the rigours of the day ahead.

Lucky dogs: Fire
Lucky cats: Benny & Spook

Aquarius

Your long-cherished dream of being in showbusiness comes a step closer this week as a square Mercury indicates that you will be asked to perform some of your old music-hall specialities as a part of a Christmas show for the 'local fallen women appeal'. Your creative use of the torch and an old Army blanket as a backdrop is sure to lend a raw authenticity to *The Pterodactyl* and *The Hindenburg* and of course the creative use of a mirror for that old favourite *Tower Bridge*. Fortunately, with no photographic evidence, the charges will be dropped.

Lucky impression: Poor
Lucky notepad: Spiral bound

Pisces

A truly marvellous week is indicated by the conjoined presence of Mars and Pluto in your fourth house. The only cloud on the horizon being a retrograde Saturn on Thursday lunchtime, which may mean a potentially dangerous attack of writer's cramp brought on by contact with a particularly sticky man from Margate. Try to avoid cowrie-shells.

Lucky tool: Tin-snips
Lucky parsley: Flat-leaf

Weekly Forecast for 5th to 11th September

Aries

Putting on a pair of overalls and touring the local office parties as 'Chuck from maintenance' is one way to satisfy your seemingly limitless appetite for free food and drink, but Venus entering your birthsign on Wednesday indicates that, as overalls are not normally worn over a pinstripe suit, there is a strong chance that your ruse may be discovered. On Friday a trine Saturn means an accident with a mince pie, the filling of which is slightly warmer than magma. Fortunately, your monkey-like shrieks and screams are captured by a BBC Wildlife unit who pay you £150.00 for a voice-over.

Lucky pan: Au chocolat
Lucky Van: Morrison

Taurus

This week, you should start planning for your retirement. Make sure you are promptly at the bank at noon to pay in the £720 of loose change you've accumulated over the years, and of course take your rightful place in the queue at the supermarket at lunchtime so you can tell the checkout girl all about your recent holiday. On Friday you will be mistaken for Archbishop Mykarios in the Post-Office. Try not to make a scene like last time.

Lucky currency: Slovenian Tolar
Lucky cartoon: Huckleberry Hound

Gemini

Unpredictable Aquarius is about to conjoin with Neptune so don't be too surprised if you don't get your trousers taken up in time for the weekend. On Wednesday, you will be taken to task by Irene from accounts over your latest expenses claim. Don't be tempted by a cut-price dibber until Mercury goes into trine on Friday, or you may regret it.

Lucky garnish: Coriander
Lucky ligament: Cruciate

Cancer

Mysterious Neptune enters your fourth house on Tuesday bringing with it a sense of foreboding about some of the domestic chores that you might have let slip recently. Regardless of how you feel about the natural world, tadpoles in your washing-up bowl is never a good sign. On Friday, Pluto's influence will begin to make its presence felt as you'll feel like plain digestives again rather than the chocolate ones.

Lucky bun: Sticky
Lucky blemish: Liver spots

Leo

Troublesome Mars enters your birthsign on Tuesday bringing to a head some of the recent arguments with your neighbours about all-night line-dancing parties. Apart from the visitors' spurs making holes in the flooring in the communal areas there is a limit to how often people want to hear the 'Dixie' air-horns from your 1987 Orange Fiat Panda with the doors welded up. A 'no account varmint' is no way to talk to the community police officer either.

Lucky Choo: Jimmy
Lucky complexion: Alabaster

Virgo

A busy week ahead with Mars conjoining with Neptune in Gemini, which, as you are only too aware, means it could be an error strewn week ahead. A mix up over King Charles and King Edward nearly causes a scene on Wednesday but disaster is luckily averted before either the chip fat gets too hot or the obedience class begins. It is also worth trying to remember the critical difference between haematite and hermaphrodite before you decide to have one polished and mounted.

Lucky tone: Dismissive
Lucky attitude: Offhand

Libra

The transit of Venus through your birthsign on Monday brings with it a sudden surge in your long-standing fascination with metamorphic rock formations. The hours of long toil in the 17-foot hole in your back garden in pursuit of this interest will yield unexpected results at the end of the week, when you discover a well-preserved Anglo-Saxon boat burial under the potting shed. At the weekend the malign influence of Pluto exerts its power when a close family member is frozen to death during a game of Bingo.

Lucky fielding position: Fine leg
Lucky particle: Quark

Scorpio

Mars forms an unusual trine with Mercury on Monday, which usually means that your otherwise sunny disposition is absent for the first part of the week. A sudden urge to buy CDs by Leonard Cohen or Coldplay would be best avoided. On Thursday, your mood will lift when your boss, who has been particularly irritating of late, suddenly decides to become a Buddhist Monk while shopping for a new electric blanket.

Lucky hinge: Concealed
Lucky flavour: Butterscotch

Sagittarius

The influence of a rising Venus will be at its peak on Wednesday when, on a whim, you decide to try a new washing detergent. The resultant allergic reaction will cause a flurry of interest from the medical profession when, what starts as a spot on your back quickly develops into a prehensile limb with three fingers. Apart from presenting you with some wardrobe issues, you will be much in demand at parties and will find fame and fortune as a multi-instrumental one-man band.

Lucky polish: Cherry Blossom
Lucky shoes: Plimsolls

Capricorn

The New Moon in Capricorn indicates that it will soon be time to start work on that dream project you have had in mind for some years. Not everyone appreciates the amount of dedication required to collect and assemble the components and create a working scale model of Tower Bridge using only your own toenail clippings. Mercury rising on Thursday indicates that inspiration for what to use for making the cables will come to you while cleaning out the plughole in the bath.

Lucky lampshade: Tiffany
Lucky language: Farsi

Aquarius

Saturn rising on Wednesday will mean you will be woken early by a Bulgarian marching band visiting your neighbourhood. This will put you in an irritable mood all morning, which will not be helped by a visit to your local grocer during which there will be a misunderstanding about whether they have peaches in stock. In the end you have to make do with peaches in syrup. The departure of Jupiter from your opposing sign early on Friday will lift your mood and by lunchtime you will have entirely forgotten Thursday's embarrassing incident with the walnut whip at the Police Station.

Lucky potato: Charlotte
Lucky lake: Ontario

Pisces

An interesting confluence between Mars and Chiron on Tuesday indicates that your long-held opinion about the collective term for eels finally gains the recognition it deserves. There will be no awards, nor all that many plaudits, but you will receive a voucher giving you a 10% discount on the power-washer you've had your eye on since the autumn of last year. Later in the week, you will be disappointed to discover that you have been betrayed by a cherished family pet.

Lucky fabric: Organza
Lucky shoes: Crepes

Weekly Forecast for 12th to 18th September

Aries

Pluto, planet of surprises, enters your seventh house on Monday.

While shopping for a new jumper, you will be surprised to discover an escaped white rhinoceros in the changing room at Marks & Spencer. Fortunately, as you know, their vision is poor and as long as you remain down-wind you will be able to change in safety. Wednesday will bring an unexpected craving for peanut butter and marmite sandwiches.

Lucky vitamin: B-Minor
Lucky flooring: Cork Tiles

Taurus

Right from the beginning of the week you find that others are interested in what you are doing, and have some good ideas to offer. It's probably best to ignore those you receive anonymously, particularly the one which suggests an illegal use of Moorhens. On Friday, you will feel a sudden urge to take up the Euphonium. Resist it, as it can only lead to heartache.

Lucky manner: Flippant
Lucky hobby: Origami

Gemini

Throughout September, everything should move into a different gear. Your ruler, Mercury, forms aspects to Jupiter and Saturn that not only swell the tide of information coming your way, but also reduce your waistline by nearly 10cm. On Thursday this week, you will be accosted in the street by evangelical atheists. Under no circumstances should you agree to attend their Bible-ignoring classes.

Lucky consonant: G
Lucky vegetable: Broad beans

Cancer

Many Cancerians love collecting, and this week may continue to turn up some real bargains. At the weekend, you may have to go as high as £15 at the local boot sale, but you will succeed in acquiring one of the missing Faberge eggs from the Russian Imperial collection. Try not to be too upset when the dog buries it.

Lucky greeting: Yo
Lucky parting: Centre

Leo

Changes at work are matched by new developments on the romantic front as Pluto transits your seventh house and indicates a brief liaison with a much older person with excessively hairy ears. Instead of hurting their feelings, why not try threading beads onto the hairs for that 'Bo Derrick look'?

Lucky strait: Gibraltar
Lucky vole: Field

Virgo

Mars entering your birthsign brings the dark clouds of conflict with it on Wednesday. It's true, that most people do bring their work home with them on a regular basis. However, last month's career change caused by a retrograde Mercury may be the exception to the rule. The animals are much better off back at the circus and the constant whip cracking and ferocious snarls have the neighbours already reaching for their phones.

Lucky treatment: Pedicure
Lucky file: Rasp

Libra

You probably felt uncomfortable with the behaviour of some of the people in your life last week. It's not every day that one finds out that the person closest to them has a secret life as a part-time Geography teacher. Despite your justifiable sense of betrayal, try to find out what led them down this dark path as it may hold some clues about the parish church, which has been missing since February.

Lucky beam: Laser
Lucky carpet: Magic

Scorpio

Things that may have been puzzling you start to get a lot clearer this week. The planet of ideas, Mercury, has some helpful exchanges with both Jupiter and Saturn, focusing your mind on something other than your troubled relationship with your prize-winning minnows. Try spending more time with them and deflect any comments about sticklebacks with a sincere apology.

Lucky sonnet: Shall I compare thee to a summer's day
Lucky bonnet: Easter

Sagittarius

Saturn rising this week indicates that you will start to question some of the fundamental tenets of your life. A trine Pluto means that you will be having a good long look at the heating bill as Autumn approaches and you may well reach the conclusion that heating your home with oil uses an average of 500gallons a year. You will begin to realise that a tenth of that amount of Vodka might do the job just as well and more cheaply.

Lucky expression: Grim
Lucky craft: Needlepoint

Capricorn

Now you can really begin to come into your own, as the move of Mars into your own sign makes other people sit up and take notice. You feel ready to face any challenge, which is just as well as on Tuesday you are drawn into a prolonged feud with the woman from accounts, the one who really thinks she's something special, despite the thread-veins. This will probably only be resolved by a duel.

Lucky weapon: Rapier
Lucky shirt: Frilly

Aquarius

You may have found that you've had to look again at things that you thought were long over. Neptune transiting your birthsign midweek indicates that you will discover a Neanderthal family living in a cave under your shed again. A more sympathetic approach is needed this time, as the traps were a disaster. Seek professional assistance.

Lucky game fish: Marlin
Lucky dance: The polka

Pisces

Many will recently have had some kind of romantic interlude with an exiled family member of the deposed President Bakiyev of Kyrgyzstan. Mercury transiting the area of your chart concerned with flings is probably to blame. Just put it down to experience and try to move on. One Friday you will accidentally swallow a hummingbird during a fire-drill at work.

Lucky element: Strontium
Lucky park: Regents

Weekly Forecast for 19th to 25th September

Aries

Routine matters take your attention this week. Everyone is still shocked by the horrors of last week. Cannibalism is quite unusual in the Home Counties, so it was bound to raise more than a passing interest when you registered the company name, You Are What You Eat Restaurants Ltd. Perhaps you should have learned the lesson from the now defunct Furry Friends Pie Company that you started two years ago.

Lucky implement: Tweezers
Lucky plug: Bath

Taurus

For all Taureans, and especially those with Taurus in the Ascendant, there's a chance that things could develop into romance this week. If you have never previously considered a romantic attachment to everyday objects such as a four-slice toaster or an ironing board, this could be just the time to do it. On Friday you will have a bad attack of whelks.

Lucky fruit: Sharon
Lucky tree: Larch

Gemini

Upsetting though recent events have been, you may begin to see the results of some of your recent actions, even those emerging from the times when you've been banging your head against a brick wall. The bandages should come off in a week or so and there was only superficial damage to the Borough Surveyor's office, so the Council will not be seeking compensation.

Lucky vision: Double
Lucky speech: Slurred

Cancer

If Cancer is your Rising Sign, you are likely to feel inclined to defend your values and your way of life this week. The more confrontational you are, the less comfortable the world around you is going to feel. You may find that you are sending out the wrong message by driving a tank to work and wearing body-armour.

Lucky spot: Black
Lucky flower: Self-raising

Leo

There is no mistaking your intentions this week. The dustbin-bag full of cotton wool, the 5-gallon can of golden syrup and the life-sized effigy of Marlon Brando have probably given the game away. A square Saturn at the end of the week indicates that your brittle toenails are starting to respond to the treatment, so the rash purchase of handcrafted replacements could have been avoided.

Lucky condiment: Sea-salt
Lucky plover: Ringed

Virgo

A magnificent aspect between Mars and Jupiter in the area of your chart concerned with wallpaper selection means that you will finally make a decision about what to put on the 'feature-wall' in the sitting room. Cork tiles are not everyone's first choice, but it does prove a practical choice when you hold your regular pin-the-tail-on-the-donkey evenings.

Lucky steppe: Russian
Lucky emission: Spurt

Libra

Usually, you are very sure of yourself – a picture of self-assurance and confidence, but this week does not see you striding out with your head held as high. This may be due to the malign influence of mysterious Neptune, which on Tuesday transits your fourth house, or it might just be the stiff neck you suffered at banger racing on Sunday.

Lucky strap: Retaining
Lucky sausage: Pork and leek

Scorpio

Everyone knows that there will be change in their lives, but you are less willing to accept this than most. Picking out any coins lower in value then 10p and throwing them in the gutter is likely to make you materially poorer in several ways, not least when you are arrested and fined for repeated littering on Thursday.

Lucky plea: Insanity
Lucky sentence: Suspended

Sagittarius

If you have been in disagreement with those around you in recent days, things may begin to improve after the mid-week trine between Mercury and Uranus when you are left a fully equipped Abrams battle tank in the will of a distant relative. However, Pluto transiting your birthsign on Thursday means that you may have trouble getting fully comprehensive insurance for it.

Lucky treatment: Holistic
Lucky goat: Snowy

Capricorn

On Tuesday, you will be involved in a drunken brawl with a blonde lady colleague during a stress-management class. However, friendly Jupiter means that you can do no wrong. When you are hauled up before the boss, he congratulates you on your vocal range and offers you both a recording contract under the name 'Atomic Mutton'.

Lucky iris: Reticulated
Lucky swan: Mute

Aquarius

On Wednesday, unpredictable Uranus enters your birthsign. You will notice a drunken, red eyed, tousle-haired lunatic beckoning at you from a shop window... You eventually realise that that's actually your reflection. An interesting aspect between Neptune and Saturn indicates that you should probably reject the long-standing offer of marriage from Jim Davidson.

Lucky trousers: Cavalry Twill slacks
Lucky herb: Marjoram

Pisces

However much you looked forward to your plans, events last week will have upset the hopes of many people with their Sun in Pisces in the last few days. On Tuesday the back of your legs will be severely slapped by a forthright lady Chiropodist with firm views on knitting patterns. Before the end of the week, you will be stopped in the street by Tiger Woods who will try to sell you a pair of shop-soiled shoe-trees.

Lucky starter: Vaseline
Lucky pudding: Bread

Weekly Forecast for 26th September to 2nd October

Aries

Mercury remains retrograde this week. On Wednesday you receive another invitation to a 'bring-your-own massage-oil' party at Kylie Minogue's house. Try to think of a better excuse than 'shopping night' for not going this time or she'll start getting suspicious and may begin to question your commitment to the relationship.

Lucky duo: Pearl Carr & Teddy Johnson
Lucky packaging: Bubble-wrap

Taurus

Your ruler, Venus, is helping you with any creative work this week. If you are involved in organising anything, especially on a large scale, the time is very auspicious. A trine Pluto means that you may be called in as a last minute body-double for Boris Johnson at the opening ceremony for the Olympic Stadium, so don't bother with the haircut you were planning at the weekend.

Lucky plane: Horizontal
Lucky joint: Elbow

Gemini

A lively aspect between Saturn and Pluto at the beginning of the week indicates that you will soon be needing new laces in those hiking-boots. A Bacchanalian lunch with a miscreant colleague on Wednesday will leave you regretting your choice of the 'fromage-a-trois' starter.

Lucky hairstyle: French-plait
Lucky wise-man: Melchior

Libra

It could be rather a low week in your love life with very few offers of marriage. Capricorn rising indicates that this might be a good time to consider giving up competitive eating. A persistent colleague with abundant warts may plague you with frequent requests for help on Wednesday. Mercury, planet of communication, suggests a bulldog-clip may provide just what you need.

Lucky wine: Gooseberry
Lucky thong: Thuperthtition

Cancer

The early part of the week may bring you an unexpected message from the past – perhaps an old flame, or possibly a tax demand. Tuesday lunchtime will see you cornered by a stocky Eskimo wearing a tie that could best be described as 'a bit sudden'. Mars rising indicates that you will narrowly lose the contest to become the new 'Face of Lentheric' to 'Lemmy' from Motorhead.

Lucky element: Surprise
Lucky shears: Edging

Leo

Saturn enters your seventh house on Thursday, bringing with it a sudden urge to take up a hobby that will help to use all those rolls of brown paper you have in the garage, and the two packets of wax crayons you found in the cupboard under the stairs. Inspiration comes as Mercury goes trine on Friday and you decide that, rather than collect autographs, you will start doing celebrity rubbings and form a club, as it can be difficult to hold down a struggling celebrity and do the rubbing, unless of course the celebrity in question is Tom Cruise or Ronnie Corbett.

Lucky knot: Sheepshank
Lucky Spice: Posh

Virgo

Pluto forms and interesting trine with Mercury this week in the area of your chart concerned with foot-care. The discolouration of your toenails will continue for the rest of the week until the growth starts to show above your shoes and attract the attentions of local scrimshaw enthusiasts who may make you an offer you were not expecting.

Lucky affliction: Restless leg
Lucky elbow: Tennis

Libra

Librans are notorious for being unable to make up their mind. This is unfair, on the whole, because you prefer to wait and look at things from every side. However, you would do well to spend your time in the company of someone much more decisive and dynamic on Wednesday evening as Mars enters your fourth house indicating that you will be attacked by a swarm of bees while queuing for a last minute birthday present in a shop that is on fire.

Lucky cake: Battenberg
Lucky table: Drop leaf

Scorpio

This is likely to be a reflective time for you, particularly since you started going through some of the old clothes you wore in the 1970s and '80s when you were a disco diva. The downside is that, as many of the shiny clothes are now a little 'snug fitting', you will be arrested on Thursday for attempting to bring down a light aircraft through careless use of a pair of silver trousers.

Lucky leotard: Black
Lucky smell: Bleach

Sagittarius

Your ruler, Jupiter, is now challenging the Sun. You are not going to take anything lying down, so it's probably not the best week for a visit to either the dentist or the chiropractor. On Friday, you will finally give up your long-standing habit of writing down the serial numbers of every banknote in your wallet.

Lucky shoes: Espadrilles
Lucky cereal: Maize

Capricorn

On Monday, your ruler Saturn is trine the Sun, and you should be able to reap the reward for past efforts. It has now been nearly three years since the safe-job and the beginning of the week is a good time to visit the scene where you stashed the loot. Avoid going on a spending spree later in the week as this may alert the authorities.

Lucky garden tool: Spade
Lucky accomplice: Big Al

Aquarius

You can't make a silk purse out of a pig's ear, so you may want to detach yourself from what's going on around you this week; but then a working holiday in an abattoir, though an unusual gift, was never going to be at the top of your list of ideal recreations. On Friday, Pluto rising means that you will have malt-loaf for elevenses.

Lucky footwear: Wellington boots
Lucky thickener: Arrowroot

Pisces

If you are part of any kind of outdoor group activities you may find yourself bending over backwards to accommodate the needs of complete strangers on Wednesday evening. A square Neptune also means that you may lose your regular parking spot when you are delayed by an earthquake in Tescos.

Lucky dog: Schnauzer
Lucky torch: Maglite

Weekly Forecast for 3rd to 9th October

Aries

Sunday's square between the Sun and your ruler, Mars, may present difficulties. Some degree of arguments or disagreements are likely, but the best way forward is to do what comes naturally to you. If lying on the floor, kicking your heels, screaming for two hours and holding your breath don't work, you could consider talking.

Lucky snack: Monster Munch
Lucky stitch: Buttonhole

Taurus

Strong aspects between the Sun and Mars, and the Sun and Uranus are going to bring some changes that you might not like. Don't let anyone pressure you into doing something that you feel instinctively is not for you. Nude skydiving is all very well, but you still have bitter memories, frostbite scars and the chafe-marks from the last time.

Lucky cat: Manx
Lucky Church: Charlotte

Gemini

Venus is now in Libra and, under this influence, romantic developments are likely. There may also be something that you've been wrestling with in recent weeks. After Tuesday's New Moon, you will see an end to both issues when you buy larger trousers and begin a torrid affair with the person you meet on the checkout desk in M&S.

Lucky match: Swan Vesta
Lucky length: Cubit

Cancer

Things will improve as your ruler, the Moon, moves into the sociable sign of Libra on Tuesday. You may receive unexpected visitors from the local operatic society in the early hours of Wednesday morning, so it may be best to make sure you have enough digestives and Jaffa-cakes on hand. Toward the end of the week, you may find yourself indecisive about yoghurt flavours.

Lucky teeth: Premolars
Lucky tongue: Aramaic

Leo

This could be quite a momentous week for you. Your ruler, the Sun, finds itself square Mars yet trine Uranus, planet of surprises. This can mean only one thing: your application to become the Vice-President of Burundi has not only been read, but actively considered. The only thing that can stand in your way now is that overdue library book.

Lucky finish: Dead-heat
Lucky cloth: Poplin

Virgo

From Tuesday onwards, new openings present themselves, and you'll be glad in a week or so if you have left behind any arguments, grudges or specialist equipment from the relationship with the village postmistress. A trine Jupiter indicates that you will become more and more interested in cattle wrangling as the year progresses, but it's probably a bit early to buy your own chaps yet. Keep renting for now.

Lucky motto: Yeee-Haww!
Lucky approach: Oblique

Libra

If Libra is rising in your chart, the new possibilities brought in by the New Moon could include some very exciting developments to do with puff pastry and the Archbishop of Canterbury. Toward the end of the week, as a special treat, you could dye that old grey cardigan purple and watch people's reactions the next time you turn up to windsurfing classes at the college.

Lucky maple: Acer Palmatum
Lucky hedge: Box

Scorpio

A passion that has been important to you up to now may cool a bit later this week. Give it time and await developments. Not everybody will understand quite how important bread-sauce has been to you over the last year. By the weekend, you will begin to appreciate that you are never happier than when you have some on, though it can get sticky under your arms and does make your shoes squelch a bit.

Lucky dish: Satellite
Lucky matting: Coconut

Sagittarius

This week, action is the key. Decide quickly, taking advantage of the tremendous astrological energy around at the start of the week, and move on to the next stage; Before you proceed, just double-check that you have adequate insurance as an escaped wolverine in the boot of your car may prove to be more difficult to remove than it was last time.

Lucky green: Lincoln
Lucky monopoly piece: The boot

Capricorn

There is a lot is going on around you and you can see clearly what needs to be done. You're in a good position to do something about it, but you may encounter stern opposition from the London Community Gospel Choir. Don't be put off now, you have already invested heavily in the wind-turbine and obtained local planning permission, so there may be enough impetus to carry the day.

Lucky shape: Rhombus
Lucky watch: Black

Aquarius

In recent weeks you've felt held back. After Tuesday's New Moon you can begin to move forward with confidence. In fact, if you have felt yourself to be blocked at every turn, Tuesday marks the moment when you realise that you have been trying to cycle with your foot-stand down. You may be bitten by a Zulu in Hyde Park on Thursday, so make sure you have the Germolene with you.

Lucky streak: The Derby & Joan Club
Lucky paste: Wallpaper

Pisces

Mercury, planet of communications, signifies that travel may feature strongly this week when you discover an old map in the drawer lining of a Victorian fighting-desk. On closer examination, the map seems to be signed A.S. which you take to be Arne Saknussemm, the man who discovered the cave leading to the centre of the earth. Try to obtain permission from Johanna Sigurðardóttirr, the Prime Minister of Iceland, before setting off.

Lucky eggs: Florentine
Lucky Pope: Pius X

Weekly Forecast for 10th to 16th October

Aries

Sunday's square between the Sun and your ruler, Mars, indicates that your feet won't touch the ground much this week. While shopping for cat-litter on Tuesday, the voluminous cloak you wear to prevent people recognising you while shopping at Aldi is caught by a freak tornado. You manage to survive on a packet of Munchies and half a bottle of lemonade before you are deposited, unharmed, outside Sainsbury's on Thursday.

Lucky polish: Duraglit
Lucky implement: Nail clippers

Taurus

Mercury turns direct early in the week, and all kinds of obstacles should just melt away. However, this will be a gradual effect and should probably not be relied on if driving at more than thirty miles per hour. If there has been some kind of legal issue troubling you, the situation will improve by the weekend when the person causing the problems drowns in a vat of tapioca.

Lucky topping: Spoon of jam
Lucky look: Smug

Gemini

Mercury, your ruler, turns direct this week, bringing you opportunities and solutions in equal measure. Whether it's making career plans, organising your personal life, or planning a comeback on the pro-celebrity jelly-wrestling circuit, this is the week to do it. By the weekend things will have returned to normal and you will develop a painful boil.

Lucky antiseptic: TCP
Lucky expression: Pained

Cancer

With a focused mind, you can manage all sorts this week. By Tuesday, you should have managed to divide a large tin into the black liquorice sticks that not many people like, the strange bobbly blue or pink ones that are just as unpopular, leaving the rest in the tin. With Halloween fast approaching, you may want to save the unpopular ones for trick-or-treaters.

Lucky gloves: Marigold
Lucky doors: Saloon

Leo

This month has not been the easiest, as hitches and mistakes seemed to crop up everywhere. Now, with Mercury turning direct, things should improve no end. By the end of the week, you will discover that a recent note to the milkman has been nominated for both a Pulitzer and Nobel Prize for Literature. It's probably a bit early to get your hopes up or to write a speech until Mars transits your birthsign later in the month.

Lucky magnet: Horseshoe
Lucky invertebrate: Slug

Virgo

This could be a difficult week. On Wednesday, you will be the victim of someone else's dishonesty or incompetence. However, once your ruler, Mercury, turns direct, on Friday, you feel much more positive and have a brighter outlook on events; besides which, you will have befriended guards and fellow prisoners alike with your nightly 'personal puppetry' shows with a torch and old blanket

Lucky suit: Striped
Lucky soap: Coal tar

Libra

This week should favour all manner of business negotiations, discussions and contracts, and you should stick to your guns. However, if your palms are still a little sweaty and causing the pearl handles on your Colt Peacemakers to slip when drawing in a hurry, try a little weightlifting resin for extra grip.

Lucky manoeuvre: Heimlich
Lucky fish: Grayling

Scorpio

If you have been waiting to move house, but encountering obstacles, the wheels should begin to turn once the bricks have been removed by mischievous children on Wednesday. You may want to wait for a tow bar to be fitted to your car rather than relying too heavily on gravity and momentum to do the work for you.

Lucky sensation: Falling
Lucky department: Accident & Emergency

Sagittarius

Usually, you make quite an impact on the world. This week, try to tread a little more lightly as the bill from the local Council for broken paving slabs is starting to mount up. You might want to consider shedding a little weight anyway, particularly as your last passport photograph had to be taken from Google Earth.

Lucky waist: Elasticated
Lucky shop: Cash & Carry

Capricorn

This is the time to seize all of the unexpected opportunities that are coming your way. Career matters should move ahead, although you need to look carefully at the small print in any offer that involves working with ferrets. On Friday, Jupiter goes trine which usually causes your ingrown toenails and Athlete's Foot to flare up.

Lucky stew: Irish
Lucky dumpling: Onion

Aquarius

Luck could be on your side in many ways, but don't commit to anything yet. Mars is about to move into your sign, and by then you'll know exactly what you think about taking on the additional responsibility of driving the Eurostar part-time while still holding down your day job. By Thursday the tests will have confirmed that it is Starlings after all.

Lucky flan: Egg and bacon
Lucky valve: Thermostatic

Pisces

Take a deep breath and nicely but firmly put your point of view to anyone who you feel has been asking too much of you, or has just been 'looking at you a bit funny'. The chances are that this week you will be understood. Benevolent Jupiter enters your seventh house on Wednesday, which means that even if it does result in a fight, you will win and there will be no witnesses.

Lucky punch: Uppercut
Lucky wine-bar: Scuffles

Weekly Forecast for 17th to 23rd October

Aries

Your ruler, Mars, is now in unpredictable Aquarius, and is about to conjoin Neptune. This just could be the month that your dreams become reality. However, it's probably best not to get too optimistic as the dreams in question may be the ones where all of your dining room furniture comes to life, grows claws and chases you down the road.

Lucky tea: Earl Grey
Lucky biscuit: Chocolate Hobnob

Taurus

The Full Moon in your sign on Thursday may be an emotional time. Just hang on in there. Keep your wits about you, particularly if Taurus is rising in your chart. On Friday, there could be some significant developments on the matrimonial front. If you are already married, you might want to consider declaring yourself in favour of a multi-partner arrangement before going any further. These things can often cause offence.

Lucky state: Utah
Lucky condition: Critical

Gemini

As Thursday's Full Moon approaches, you may feel irritable and on edge, but in many ways you can expect some positive experiences in the next week or so, especially if you are as flexible as your sign, and the graffiti on the staff toilet wall, suggest. At the end of the week you will have an accident with a bacon-slicer when he backs into you in the supermarket car park.

Lucky marsupial: Opossum
Lucky wave: Dismissive

Cancer

All those with Cancer strong in their chart may experience some major changes in their lives at the end of this week. Especially if you have Cancer rising, you could find that your duties and obligations change in a fundamental way when you decide to become a hermit and live in a single stone cell on a remote Scottish island. On a positive note, you have always enjoyed fresh air and porridge.

Lucky shirt: Hair
Lucky look: Unkempt

Leo

Mars has moved into your opposite sign which generally means trouble ahead. You will have no trouble standing up for your point of view, but exercise diplomacy, particularly if there are arguments with your nearest and dearest. This is rather an unpredictable week for you, and it will be best if you can keep your arguments to times when sharp knives, frying pans and boiling water are not within easy reach.

Lucky position: Ducking and diving
Lucky suit: Plate armour

Virgo

This week sees you set on a new and interesting course: Footwear Through The Ages at the local college. Uranus promises unexpected developments toward the end of the week, particularly for anyone already involved in a long-distance relationship with a one-eyed cross-dressing ex-Mafia hit man called Margaret Weaver.

Lucky sauce: Béarnaise
Lucky Sea: Sargasso

Libra

There could be new acquaintances, new social opportunities or even a new love in your life by the end of this week. Your ruler Venus is in company with Mercury for a while and you will delight your friends with your wit, repartee and skill with stop-action animation. An unexpected phone call or letter on Thursday will bring good news about your appointment for prosthetic eyebrows.

Lucky vitamin: B-Flat
Lucky game: Pooh-Sticks

Scorpio

On Friday, your ruler Pluto is in opposition to Saturn. Only a day or so later, Jupiter turns retrograde. This can only have one possible outcome. Your usual practice of trying to get down the stairs before the flush finishes is all very well at home, but at work the stairs are longer and the flush shorter, so sooner or later the monster will catch you.

Lucky trousers: Purple loon-pants
Lucky headgear: Turban

Sagittarius

Many people with Sagittarius strong in their charts feel they are having rather a rough time at present. This is hardly surprising given the presence of Mars and Neptune in the area of your chart that concerns relationships. The tensions should have eased somewhat by Wednesday when the restraining order comes into force and the 24-hour security guards are in place.

Lucky precaution: Panic room
Lucky attitude: Watchful

Capricorn

This week could be very exciting and creative, as you lay down some serious foundations. Later in the week, a visit from the local CID is indicated as they are still concerned about the suspicious disappearance of Amway resellers in your area and want to check whether the concrete has set. Try to remain calm and, if possible, hide the 2,156 bottles of carpet-cleaner.

Lucky soup: Primordial
Lucky snack: Cheese on toast

Aquarius

The pace of your life is going to leave the rest of us breathless. Mars has moved into your sign and, although you have the drive and energy you need, you should watch out for disagreements with the man from number 47 as, unbeknown to you, he has an extensive collection of edged weapons and a mind unhinged from years of working in the local tax office.

Lucky steps: Rapid
Lucky number: 999

Pisces

Mars meets your ruler, Neptune, at the end of this week. You may feel restless and confused. It's not really the best time to act, so you may want to reconsider your decision to take on the lead role in the local drama society's lunchtime production of *Boadicea* at the working men's club. On Thursday, though your usual meat-pie will be tasteless and soggy, it is generally not a good idea to rub it into the hair of the person serving you.

Lucky publication: *Home & Garden*
Lucky section: Lost and Found

Weekly Forecast for 24th to 30th October

Aries

November could be one of your best months for some time. However, your ruler, Mars, is now in unpredictable Aquarius and is about to conjoin Neptune, so any rash decisions you make about floor coverings for the bathroom are likely to come back to haunt you later in the month. On Wednesday, someone will push a ginger biscuit through your letterbox bearing the message in green ink, 'Do not eat this biscuit as green ink is bad for you'.

Lucky mirror: Convex
Lucky ride: Nemesis

Taurus

The Full Moon in your sign on Thursday means that you may be cautious and hesitant, but be prepared to think on your feet as on Tuesday lunchtime, while you are out with a client, thieves break into your office and steal all of the chairs. On Friday, you suffer a mild concussion as you are struck a glancing blow with a frozen chicken in the British Library reading room.

Lucky agenda: Secret
Lucky moth: Hawk

Gemini

There are likely to be some major changes in your life during November. They could come as a surprise right at the beginning of the month, as you discover a secret passage from your conservatory to the top floor of the Borough Council Highways Department. On Thursday, you would do well to avoid anyone with one leg and a Welsh accent.

Lucky paint: Magnolia
Lucky vegetable: Cauliflower

Cancer

Many Cancerians are feeling very tense right now. You may be among the many who will find the Full Moon on Thursday particularly trying. However, the end of the week brings the light-relief you so badly need as it is Halloween and you can look forward to the little faces of the 'trick or treaters' on the doorstep lighting up when you hand out your wasabi profiteroles and toffee-onions.

Lucky ticket: Cheap day return
Lucky disguise: Hank Marvin

Leo

This will be far from a quiet week. Mars has moved into your opposite sign, and right from the beginning there could be arguments and conflicts around you, some of which could become rather heated. Admittedly it was cheap, but you may live to regret your choice of a budget wine-tasting tour of Kabul.

Lucky precaution: Kevlar vest
Lucky trousers: Brown corduroy

Virgo

If Virgo is rising in your chart, the emphasis will be on practical matters. Your methodical approach and steady hand will see you reach the quarterfinals of the World Pro-Celebrity Jenga championship. However, Pluto goes trine by the last day of the tournament and you develop a nervous twitch, a heavy head cold and a racking cough.

Lucky decongestant: Sudafed
Lucky sleeve: Left

Libra

You may fare well this week, even if there are some changes in store for you by the weekend. You will have a heated disagreement with a shop assistant on Wednesday. A misunderstanding about the possible use of bag-charms when asking about changing-room facilities may lead to an unseemly scuffle with the security staff. As Mercury will be in the ascendant by then, no charges will be pressed.

Lucky plug: 13amp
Lucky fuse: Short

Scorpio

The week may open with some kind of domestic conflict and you will find it useful to be prepared for the possibility. Keep your head down, chin tucked well into your chest, and lead with your weaker hand. You have always been susceptible to a left-feint and right-cross combination, so keep focused on your footwork and circle to the right where possible.

Lucky round: Three
Lucky trunks: Stars and Stripes

Sagittarius

This could be an odd sort of week. On the one hand, old issues from the past may surface, not to haunt you but to give you a chance to put things right, or maybe just to learn from past mistakes, such as the incident in the cinema when you were trying out your new clip-on trousers. But the less said about that, the better.

Lucky movement: Jewelled
Lucky hand: Second

Capricorn

There could be distinctly difficult times in the financial markets. Capricorn people may be harder hit than most but, being fore-warned as you are, you are able to turn a very nice profit on a pair of 17th century jewelled duck-mounts that you have had put by for some years. On Thursday, a trine Venus indicates buttered crumpets for elevenses.

Lucky look: Retro
Lucky taste: Salty

Aquarius

On Thursday, you may find yourself looking after friends in trouble or at least in need of sympathy. Next weekend may bring some extraordinary changes to the structure of your life when you and your current partner are accidentally elected 'Carnival King & Queen' of Stepney. As neither of you have been to Stepney, quite how this happened will remain a mystery.

Lucky walk: Lambeth
Lucky lunch: Jellied eels

Pisces

Talk things over with a trusted friend, or just ride the wave of uncertainty for a few days. Communications may be an issue this week. On Wednesday, while mowing the lawn, you will receive an unexpected visit from the rather attractive lady at number 14 wanting to 'try something out'. Your enthusiastic agreement wanes a little when you find out that what she is keen to try is a can of Mace.

Lucky top-dressing: Shredded bark
Lucky hoe: Dutch

Weekly Forecast for 31st October to 6th November

Aries

Normally the most decisive of people, you will be plagued by bouts of uncertainty about your actions on Monday, when you complain loudly about the pensioners clogging up the queue at the supermarket at lunchtime. On the way home, you will be accosted by a rowdy group of elderly men and women, dressed in 'colours' (grey and beige) who are in the process of draping carpet-slippers joined with a length of string over the overhead telephone wires.

Lucky precaution: Bulletproof string vest
Lucky publication: *The People's Friend*

Taurus

Tuesday's New Moon in your opposite sign means that this would be a good day to seek advice on an important matter relating to zebra finches. Some of the minor details may seem unimportant when you first look at them, but pay close attention as Mars goes direct on Wednesday and anything which may counteract the effect of the lizards can only be beneficial.

Lucky insect: Hover-fly
Lucky spring: Transverse-leaf

Gemini

In recent months you have been tied down much more than you normally like. This is about to change, when Mars goes direct midweek and you discover the secret diary of Harry Houdini concealed inside a life-sized papier-mâché model of Sir Alec Douglas Home. On Friday you will meet a man who will try to sell you a shop-soiled BMX Wheelbarrow. If you are canny, you might get a bargain.

Lucky adhesive: Superglue
Lucky confection: Marshmallows

Cancer

On Tuesday, Pluto forms an unusual square with your ruler, the Moon, in the area of your chart concerned with health and wellbeing. You have recently lost so much weight that on Friday, your friends, concerned for your health, will club together to pay for you to have your jaws wired open. It's probably not the best weekend to go swimming or have a passport photo taken.

Lucky additive: Monosodium Glutamate
Lucky file: Ring binder

Leo

Misunderstandings are sufficiently likely this week to suggest that you tread very carefully, as on Thursday afternoon you may well find yourself facing a term in a foreign prison for trying to sell relabelled tinned spaghetti into Saudi-Arabia as Arabic alphabetti-spaghetti. The plan is a sound one except that the first sample you hand out says something rude about King Abdullah's pet rabbit.

Lucky diet: Bread and water
Lucky chains: Manacles

Virgo

Always a worrier, you may be too hard on yourself. Not everything is your fault. On Tuesday, a birthday present to a little girl will cause considerable offence. You were asked for something to go with a Nurse's uniform – a 'Barbie Hospice' was probably not what the parent had in mind, though not as bad as the 'My Little Pony Glue-Factory' you bought at Christmas.

Lucky escape: Bathroom window
Lucky excuse: Whiplash

Libra

Although Mercury and Venus are travelling together in your sign, providing a congenial backdrop to your life, each planet has a difficult aspect to Neptune early in the week. This means that your application to become a major saliva donor for the local hospital trust is turned down and the 'free samples' you provided are sent to the Crown Prosecution Service as evidence.

Lucky motorway: M26
Lucky services: Knutsford

Scorpio

Mysterious Neptune enters your birthsign on Thursday, bringing with it, renewed concern about the damp-patch on the wall of the dining room. When you get someone in to look at it, investigations show that it is down to a mains leak under the floor. It has obviously been a problem for some time, as there is now a thriving coral reef.

Lucky bell: Diving
Lucky accessory: Aqualung

Sagittarius

Recent events have left you even more philosophical and reflective than usual. Your altruistic nature and willingness to help others have always been both a blessing and a curse. On Tuesday, you will be assaulted in the street by a group of partially sighted football supporters who, after shaking your hand in greeting, read the fine collection of warts on the back of your right hand as something rude about their team in Braille.

Lucky expression: Resignation
Lucky pants: Lurex thong

Capricorn

Money is going to be at the forefront of your mind throughout November. Your recent attempt to become a hip-hop MC was something of a disaster when you made the observation 'How hard can mucking about with record players actually be?' On Friday, your prowess with the Bassoon is called into question in the Co-Op. Try to remain calm.

Lucky shape: Lozenge
Lucky lozenge: Throaties

Aquarius

Your creative powers are at an all-time high. However,

bombarding your local railway franchise's Marketing Department ('Ideas Above Our Station') with suggestions for railway themed meals from the buffet mostly fall on stony ground, with the exception of 'Beef Encounter', which wins you a free return ticket to Stratford.

Lucky Tray: Milk
Lucky Knife: Mack, the

Pisces

Piscean's care a lot about others, but can get into terrible muddles themselves as your large collection of badly made 'Guys' acquired for a penny each proves beyond reasonable doubt. On Friday, a nine-hour siege ensues when a neighbour spots a stack of what he takes to be dead bodies on your compost heap. It is not until you discover that the people outside your house are the police, rather than Avon representatives, that you give yourself up.

Lucky lotion: Baby
Lucky baby: Jelly

Weekly Forecast for 7th to 13th November

Aries

Mars, your ruler, is now in your own sign for the first time for a couple of years, and your life is really going to take off, provided that you can do something about the pronounced lisp you seem to have developed since the accident with the cocktail sausages. On a more positive note, your hot duck soufflé at the boot sale refreshments stall is nothing but a triumph.

Lucky plate: Pewter
Lucky brassiere: Black lace

Taurus

For you, the accent is on work, career and business. In the early days of this week, you will be introduced to a visiting trade delegation from Holland. Although you will do your best to please, constant referrals to Holland as the Neanderthals will start to grate on them by the end of the week. The correct way of referring to their country is 'The Nether Regions'.

Lucky cheese: Gouda
Lucky infestation: Mice in clogs

Gemini

An international incident is narrowly averted on Tuesday when Mars conjoins with Saturn in the area of your chart concerned with communication. A once in a lifetime opportunity to join the Swedish mixed softball team for an after-match sauna goes horribly wrong. As it is your first sauna, you completely misunderstand the request to 'Pass water for the stones' and leave under something of a cloud.

Lucky footwear: Flip-flops
Lucky towel: Ivor the Engine

Cancer

Your ruler, the Moon, goes retrograde on Tuesday which indicates that the end of the week could have some surprises in store. On Thursday and Friday, there will be a romantic interlude with the band of the Welsh Guards at your local betting shop, which may not be anything permanent, so you should watch out that they don't become more possessive than you can comfortably handle.

Lucky tune: Cwm Rhondda
Lucky round: 275 pints of Brains bitter and a pickled egg

Leo

Mercury rising means that there is a good chance that a new career path will open up toward the end of the week when you are dismissed from your existing position for experimenting with Alchemy in the executive washroom at lunchtime. Now is the time to prepare yourself, and on Thursday and Friday you will really have an opportunity to shine. Next week, you will be allowed to apply the polish and cover up scuffmarks on the heels and toes.

Lucky accessory: Suede brush
Lucky laces: Waxed cotton

Virgo

You love to be useful, and this month you realise that you truly are. Fortunately, people with Virgo prominent in their charts often have unusually prehensile extremities, which can be pressed into service in tricky situations, or to distract a fractious child. On Tuesday, a close family pet will be carried off by a golden eagle.

Lucky precaution: Fine netting
Lucky racquet: Tennis

Libra

People with Libra strong in their charts are good communicators and can explain things very clearly, especially on Thursday when Mercury enters your birthsign and you are required to explain the presence of a receipt for dinner for two from a smart restaurant in Paris, when you were supposed to be on a team-building course in Ulan Bator.

Lucky explanation: Plausible
Lucky racket: Deafening

Scorpio

Life has dealt you a different timetable this week. Go with it, and you'll find that unexpected good things await you. If you resist, you deny those things to yourself and run the risk of being expertly isolated from your protection squad, kidnapped and held to ransom by the goats again. The good news is that they are just as happy with newspaper as they are bundles of used currency.

Lucky pencil: HB
Lucky rubber: Latex

Sagittarius

Love could be in the air. In fact, if you begin a love affair at this time, it could feel as though this is the one you have been waiting for all your life. Sometimes such a feeling is no more than idealised infatuation, but this time, with Saturn trine both Neptune and Venus, it could be something really lasting, that doesn't involve specialist equipment, a trapeze or planning permission.

Lucky confection: Ruffle-Bar
Lucky infection: Pyorrhoea

Capricorn

With Mercury retrograde, you will feel a need to reconsider some decisions that you made earlier in the month. Mars in

Aries is putting a new boost of energy in your life right no, and you can achieve a great deal – if you can take a week or two off from cage-fighting classes and devote your energies to your real vocation: breeding and training stunt hamsters.

Lucky glands: Lymph
Lucky pores: Closed

Aquarius

Your ruler Uranus is preparing for an argument with the quarrelsome Mars, and it seems inevitable that sparks will fly. Nevertheless, you are going to need to take part in some team-work and, despite your best efforts, will get into at least one fight and a minor scuffle over a disagreement about WC Fields and whether he was magnetic or arable.

Lucky town: Philadelphia
Lucky mushroom: Chanterelle

Pisces

On Saturday, Jupiter goes direct after its long retrograde period and this will get things moving firmly to your advantage by the beginning of the week. On Tuesday, when cleaning out your garage, you discover the Golden Apples of the Hesperides in an old shoe box and get into a heated argument with Mr. Heracles from the greengrocer about whether they are Granny Myths or Golden Fictitious.

Lucky titan: Prometheus
Lucky advisor: Athena

Weekly Forecast for 21st to 27th November

Aries

Mercury, planet of communications goes retrograde on Tuesday which means that your forthcoming trip abroad may be more troublesome than expected. However, there is some good news: the book you bought about local customs turns out not to be, as you thought, a guide to the habits of the natives of the country, but a useful manual on how to negotiate passport control and immigration.

Lucky queue: EU Passports
Lucky manner: Obsequious

Taurus

Mercury is currently travelling backwards through your sign at the moment, which indicates there will be some serious misunderstandings this week. On Thursday, you will have to go home to get changed after a misprint on the invitation to the 'Crosse & Blackwell Salad Dressing Competition' in which the words '& Blackwell Salad' were omitted.

Lucky tube journey: Short
Lucky weather: Fog

Gemini

Venus is now in Libra, so romantic developments will be likely for some people with Gemini strong in their charts – particularly for those who have invested heavily in equipment. An older colleague, perhaps a close friend, may offer you valuable advice on metal fatigue. They will be wise words – heed them well.

Lucky pint: Light and bitter
Lucky patois: Jive

Cancer

An interesting aspect between your ruler, the Moon, and Mars means that there will be conflict at work with someone in a position of authority. If you could avoid referring to them as 'an unrefined nose-picking yahoo', you will find your future employment prospects significantly enhanced.

Lucky position: Nose to the grindstone
Lucky lips: Sealed

Leo

On Wednesday, the malign influence of Chiron is felt most strongly by those ruled by the Sun. When you pop out to lunch you will accidentally spill the pint of someone so large that they have their own postcode. The good news is that as you are still fairly spry, you can flee the premises before the looming shadow engulfs you and not have to resort to begging for mercy while wetting yourself, as usual.

Lucky steps: Long
Lucky hormone: Adrenaline

Virgo

On the way to the office on Thursday morning, you will be delayed by a road closure caused by a minor collision between a moped and a parked car. Unfortunately, the incident will be seen as a perfect opportunity for the Motorway Police to conduct vital research on a government initiative to find out whether natural or man-made fibre brooms provide the best means of clearing broken glass from the public motorway. Do not over-react as you may put the officers off their picnic.

Lucky virtue: Patience
Lucky language: Strong

Libra

After nearly three years spent under the influence of 'Saturn-Uranus', finally there is some movement in the area of your chart concerning Country Music. So it is probably best to unpack your chaps and Stetson, as you will be invited to participate in the Smokey Mountain ten-mile sponsored 'Mosey' in Lincoln.

Lucky appliance: The Goblin House-Maid
Lucky pause: Pregnant

Scorpio

Mercury goes retrograde in the area of your chart concerned with cakes and biscuits on Thursday, which means that it's more likely than not that the canteen will have run out of cinnamon buns before you get there for elevenses. On Friday you will overhear a plot to overthrow your Chief Executive and replace him with a sack of potatoes. You need not concern yourself as this has happened twice in the last three years and no-one noticed.

Lucky pen: Fountain
Lucky reflex: Yawn

Sagittarius

There could be some disruption at home this week. You may find yourself inundated with visitors from Minsk, and then discover that a fuse blows or there's a flood in the bathroom. The good thing is that you are at your most inventive and can rise to the occasion with aplomb as you received a matching pair of plombs for your wedding anniversary.

Lucky instrument: Dulcimer
Lucky elbow: Right

Capricorn

An unusual aspect between Venus and Pluto will make its presence felt on Wednesday when your recent hobby involving wild mice and correcting fluid comes to light. Be warned, the animal welfare people are very persistent and even though you have been using a trap that does not harm the mice, painting white stripes on them with Tippex to make naturalists think that they have discovered miniature badgers could be seen as animal cruelty.

Lucky cat: Corky
Lucky treatment: Worms

Aquarius

Sleep will continue to be a problem this week with Mars entering your seventh house, dropping his change on the landing and tripping over the mat in the bathroom. Fortunately, by Wednesday this noise will be drowned out by the usual sirens and car-doors slamming until 3am, so you can expect to catch up then.

Lucky metal: Bismuth
Lucky instrument: Zither

Pisces

On Tuesday, an unusual aspect between Neptune and Jupiter indicates that a defect in your recently purchased microwave means you are transmitting your food rather than heating it up. However, by careful adjustment, you discover that you can actually receive other people's dinners instead of buying your own. On Friday, you will have an unruly mob of TV chefs on your doorstep. Try to remain calm.

Lucky bandwidth: 25.1MHz
Lucky gravy: Onion

Weekly Forecast for 28th November to 4th December

Aries

Your natural bonhomie can sometimes get you into unexpected scrapes from which it can be difficult to extricate yourself. Monday evening's usual quiet drink at the local pub turns into something of a wild-west-style brawl after an offhand remark kicks off a violent disagreement with the vicar over whether the Earth's crust puts hairs on your chest.

Lucky window: Lounge bar
Lucky manoeuvre: Duck

Taurus

Tuesday may be a trying day for anyone with Taurus rising. What you thought would be a simple errand becomes something of a chore after you mishear the original request and spend all lunchtime looking for a pattern for a knitted barracuda. Communication problems continue to plague you when you overhear a whispered conversation between the Mayoress and the Bishop during which she admits to preferring the taste of the Coleman's, without realising that the topic being discussed is mustard.

Lucky scrub: Apricot facial wash
Lucky exfoliant: Gravel

Gemini

Mercury's influence is felt most strongly today when a misdelivered internal memo provides a glimpse of the tormented, horizonless landscape that is the HR department. On a brighter note, you will receive some news for which you've waited a very long time – you are to get your own card in the company 'Top Trumps' game.

Lucky markings: Piebald
Lucky tree: London Plane

Cancer

Your peculiarly large consumption of industrial protective clothing is finally your undoing this week. It was only a matter of time before someone put two and two together and revealed that you have been travelling the country for the last ten years in secret and placing a single red rubber workman's glove on every roundabout in the country. On Friday, seed-cake will be on special offer in the Co-Op.

Lucky plea: Guilty as charged
Lucky fit: Coughing

Leo

An unusual trine between your ruler, the Sun, and Uranus means that there is likely to be a bit of a scene at the doctor's surgery on Wednesday when the takeaway delivery driver is admitted to hospital after the doctor heard him mention that he had both Chinese thighs and Cumberland ring.

Lucky layer: Troposphere
Lucky wind: Prevailing Westerly

Virgo

Mars entering your birthsign this week usually leads to an overwhelming urge to do things differently. By all means try some of them out, but make sure you have warmed-up properly first – you know what your back is like. On Friday, you will discover half-a-kilo of top grade pigs liver in your sock drawer.

Lucky eye: Private
Lucky pill: Ibuprofen

Libra

At the beginning of the week, you will have a sudden urge to do something completely different and on a whim fly to Marseilles to join the French Foreign Legion. The good news is that you will be talked out of enlisting by an unlicensed bullion dealer during a visit to an all-night eyebrow bar, at which you are having an additional pair fitted.

Lucky cap: Breton
Lucky gesture: Gallic shrug

Scorpio

The focus is now on practical matters. If you are looking for more money or a promotion, the chances are better than they have been for some time. Your recent decision to not wear your fishing hat and waders in the office anymore will have been noted. Not spinning round on your office chair until you are so giddy that you crash into other people's cubicles has also made a positive impression.

Lucky melon: Honeydew
Lucky adhesive: Pritt

Sagittarius

Mars is square your ruler, Jupiter, making sure that, whatever else, this week will not be boring. In fact, things may get just a little too hot on Monday when a disgruntled former employee runs amok in the accounts department with a flame-thrower. Luckily for you, the Finance Director, who has always had a reputation of something of a wet blanket, proves his true worth.

Lucky extinguisher: CO2
Lucky escape: Fire

Capricorn

You have particular plans and have been working towards them diligently without much outward sign of success. If you lose your cool on Monday, don't despair because all kinds of unexpected and exciting things are waiting for you just around the corner. On Thursday, a vagrant in a duffle coat and stilettos will press an outstanding recipe for date slices into your hand as you go to lunch.

Lucky feature: Hazard lights
Lucky taxi driver: Brian

Aquarius

Some Aquarians found romance last week and for some of you it has seemed that at last you have met your soul-mate. This week reality sets in when you discover the grim truth that the person you took to be your perfect match was actually only interested in getting hold of your impressive collection of antique prosthetic limbs.

Lucky flame: Old
Lucky firearm: Brown-Bess

Pisces

Mercury is retrograde in your opposite sign, and not everything is clear to you right now, but Sunday's aspect between your ruler, Neptune, and Mercury have enabled you to understand some things better.

Lucky lift: Fireman's
Lucky bow: Recurved

Weekly Forecast for 5th to 11th December

Aries

Mars enters your fourth house on Wednesday, indicating that the new HR Director has more draconian ideas about qualifying for staff loyalty bonuses than their predecessor. Although the Marketing department is initially happy about the prospect of new branding, it's not until the charcoal brazier and irons make an appearance that people begin to catch on.

Lucky colour: Burnt Umber
Lucky bridge: Sydney Harbour

Taurus

On Monday, Pluto goes retrograde in your opposite sign. This can only mean one thing: you will finally find someone to teach you the rudiments of the closely-guarded secret Japanese Coffee Ceremony 'Kenco' during which people make a really nice cup of fresh coffee while defending themselves against men in armour who are trying to hit them with sticks.

Lucky spoon: Dessert
Lucky kettle: Electric

Gemini

An interesting aspect between Jupiter and Saturn on Wednesday indicates that you will have a violent argument with the local delicatessen owner over what he actually meant by the words 'would you like to try my sausage?' whispered to your partner in the waiting room at the bus-station. The words 'not with a cheese-knife you don't' will gain a new and sinister significance.

Lucky sausage: Chorizo
Lucky olives: Kalamata

Cancer

The start of this week may bring you some significant opportunities, as a fabulous aspect between the Sun and Saturn accompanies the Full Moon, your ruler. This couldn't happen at a better time for you, with Mars filling you with ideas for the future and giving you the energy to try the break-dancing moves you've been practising in front of the bedroom mirror before the office party next week.

Lucky lamp: Lava
Lucky Psalm: XLV

Leo

On Monday, the Leo Full Moon could see you falling head over heels in love, or in some other way it could be a very emotional time. When it appears, your initial excitement means you could come over altruistic. You may very well be carried away – possibly an alien abduction – but it's much more likely to be Pixies again.

Lucky guess: Plutarch
Lucky garnish: Chives

Virgo

On Friday, the Sun challenges your ruler and the interesting asteroid known as Chiron in a celestial position known as the 'wheelbarrow'. Don't look, it will only upset you. The weekend will bring good news about your new gazebo. There may well be a misunderstanding with the man who delivers it when he asks if you want decking.

Lucky force: Air
Lucky soup: Carrot and coriander

Libra

Never the most decisive of people. Pluto enters your seventh house this week and with it comes one of the most difficult dilemmas you have ever faced. Ultimately, your choice will decide the outcome of the ever-sensitive peace talks between the Barry Manilow fan club and the Ministry of Food and Fisheries. A roll of greaseproof paper will prove even more useful than you could have imagined.

Lucky snack: Anchovies
Lucky port: Tawny

Scorpio

An offhand remark could lead to problems on Wednesday. It had become accepted practice to poke fun at the people who visit your local pub to sell copies of the *War Cry* magazine. What you nay not know is that the local citadel has been taken over by the Militant Wing of the Salvation Arms, who's Christianity is an altogether more muscular brand. On a positive note, the ambourine will still be playable after removal.

Lucky hymn: 152 – *Fight the good fight*
Lucky cushion: Soft

Sagittarius

In recent weeks you have been tied up much more than you normally enjoy. If you've been feeling constrained, you will begin to shake off any fetters and start to feel a real sense of freedom, so it may well be worth reconsidering renewing your monthly membership to Tracy's Dungeon.

Lucky chain: Bicycle
Lucky lock: Teddington

Capricorn

The Sun moving into your birthsign brings with it a sense of serenity you have not experienced for many months. It seems that nothing can upset your feeling of wellbeing until Tuesday when a close relative is injured by a falling tree during a flight to Bucharest. On Friday the heel will come off of your best dancing shoes during a vigorous gavotte.

Lucky yoghurt: Rhubarb
Lucky block: Breeze

Aquarius

Contract negotiations are likely be to the fore this week. On Monday, you will be aware that you are getting some funny looks during a particularly intense negotiation session after you have been chewing the end of your red-pen. Following the meeting, you will discover that the pen has leaked, leaving you looking like a post-rampage Hannibal Lecter.

Lucky attitude: Devil-may-care
Lucky solvent: Nail polish remover

Pisces

This week, you are starting to give some thought to what to buy your nearest and dearest for Christmas. When a remark is casually dropped into the conversation about 'getting something romantic to light up an evening', scented mood-candles would probably be a better received gift than a two-cylinder diesel, ten-horsepower, mobile lighting generator and gantry on a stainless steel trailer.

Lucky stance: Defensive
Lucky mood: Contrite

Weekly Forecast for 12th to 18th December

Aries

Mars brings a new sense of energy and dynamism to Aries this week. However, a little caution should be exercised at the staff Christmas party on Friday when a particularly lively impersonation of Pete Townshend may leave you with a nasty case of 'Air Guitar Elbow', which if it not treated properly may spread and leave you looking like Iggy Pop.

Lucky bandage: Triangular
Lucky evidence: Photographic

TAURUS

You're in a creative mood. If at all possible, you are not interested in daydreams. Putting ideas into practice is your way, and this week should see you well on the way to success with one practical and imaginative project. Don't be too impatient with it; a scale-model of the Home-Secretary made out of Meccano is a tough and unrewarding job.

Lucky pill: Iron
Lucky plaster: Waterproof

GEMINI

If you are involved in any kind of joint venture at the minute, you should stick to this principle even more resolutely. There is a good market for hip-replacements right now, and even though you have no experience in orthopaedic surgery, your can-do attitude will win you a lot of admirers.

Lucky insurance: Professional indemnity
Lucky premises: Temporary

CANCER

The influence of Uranus fills the air with a sense of foreboding over the next few days. Consider changing your curtains before the worst happens. Wednesday's aspect between Venus and Neptune will probably mean that you will have a misunderstanding involving brown paper at the post-office. A man with a sticky handshake and a twitch may try to sell you a vampire costume.

Lucky tipple: Calvados
Lucky container: Tupperware

Leo

You are entering a very favourable time. Whether you are involved in work, domestic matters, or learning and education, this week gives you the chance top put yourself right on top of a situation when you join an acrobatic act at a touring circus. A leotard fitting may go horribly wrong on Wednesday, but nature, in her wisdom, provided you with two of the most vital things, so losing one will not be the blow it might have been.

Lucky coating: Artex
Lucky fluff: Navel

Virgo

Wednesday and Thursday will be days when you can forge ahead with anything that you are doing, and you will probably do it in the company of friends or representatives from the Department for Work and Pensions. If there is anyone special in your life, Friday could be a significant day as the restraining order is due to expire.

Lucky muscle: Triceps
Lucky plant: Chemical

Libra

One problem common to many people with Libra strong in their chart is a tendency to put off till tomorrow what they know they really ought to do today. Wednesday is no exception to this when you are close to completing a particularly challenging jigsaw and the chip-pan catches fire. It is probably best to leave finding the horse's head from Constable's *The Hay Wain* until after the fire is out.

Lucky precaution: Damp tea towel
Lucky guess: Still in the box

Scorpio

You can work towards your goals with some success this week. Family and friends are more important to you than you let on, but just now you need to be careful, because those around you are likely to be irritable and easily upset by your late-night 'one-man-band' practice sessions. On Thursday, you will be accidentally run over by an unmanned motorised wheelchair.

Lucky paper-size: Quarto
Lucky mug: Enamel

Sagittarius

You have a number of obligations at the moment involving your work and your immediate finances. You tend to keep a cool head and on the whole manage everything very efficiently. But be careful not to neglect your home and your relationships, as Mars rising indicates that all of your family, friends and colleagues are seriously considering emigration.

Lucky pasta: Linguini
Lucky sauce: Carbonara

Capricorn

Money matters are to the fore this week. Your recent economy drive has not left everyone as happy as they might be. On Thursday, the damp patch in the hall gets so bad that there is a permanent rainbow when the light is on. In an attempt to dry it out, you borrow an old paraffin heater that manages to transform the cold damp hall into a cold, damp hall filled with paraffin fumes.

Lucky shout: Exasperation
Lucky throw: Over the garden fence

Aquarius

Your diet has not been quite as healthy as it should have been of late. The forthcoming festivities are unlikely to improve the situation. You are already known as something of a 'nutritional over-achiever' and the buttons on your shirts have now started to resemble a burst sausage. There is good news around the corner. A carelessly defrosted turkey will give you typhoid and you will lose forty pounds in hospital.

Lucky convulsion: Dry heaves
Lucky state: Febrile

Pisces

With Jupiter travelling through Aries you can expect a hilarious incident with a tin of gloss-paint at the cinema that changes your fortunes by mid-week. On Thursday, Moira Stewart breaks into your house and hides nectarines behind the cushions. Your need for freedom will come to a head on Saturday when, on a whim, you will find yourself buying a one-way plane ticket to Trondheim.

Lucky approach: Westerly
Lucky sea-area: German Bight

Weekly Forecast for 19th to 25th December

Aries

As a Fire sign, you are normally quite at home with all aspects of cooking, but icy Neptune transits the area of your chart concerned with food preparation and casts doubt over the advice you have been given for defrosting the turkey. It will normally be properly defrosted in two days, rather than what you have been told: 'When it has cobwebs in it'.

Lucky stuffing: Applc and thyme
Lucky medication: Imodium

Taurus

This week will be great on the whole for getting group projects off the ground. You may find that people in your community are looking to you to do something for the general good. You may gain considerable local support for a 'Torch 'n' Pitchfork' march to the offices of the Borough Council, but they will be closed for the Christmas Holiday and the few people you do get interested will be waylaid by Carol singers.

Lucky coat: Duffle
Lucky mittens: On strings

Gemini

This is not a week to rush into things. That isn't so easy for Gemini, for no sooner have you got an idea than you are rushing to put it into action. The one about dressing up as a dolphin and covering yourself with greengage jelly during the office party was not a particularly good one and you can expect to be carpeted by the end of the week. The good news is that the carpet is an antique Persian rug so the burns are minor.

Lucky herb: Miller
Lucky grass: Fescue

Cancer

The planets are setting themselves up to give you every opportunity to get on in every part of your life. If you have special ambitions, you are likely to find the opportunity to move them on very significantly. Your finances are likely to improve after you find £200,000 in used £10 notes down the back of the fridge while trying to retrieve an escaped duckling.

Lucky spree: Spending
Lucky look: Supercilious

Leo

Your ruler the Sun is in dispute with Pluto this week. Although you may be having arguments with someone close to you, this will be a very good month in which to advance your career. However, not everything will go smoothly after the resignation note allegedly from your boss is found to be a forgery and a detailed forensic examination of your car finds him chained to the engine with tape over his mouth.

Lucky gland: Pituitary
Lucky tape: Gaffer

Virgo

Mars enters your fourth house on Tuesday bringing a potential catastrophe with it. Your power will go off at around mid-morning threatening your huge collection of frozen commemorative egg and cress sandwiches from significant dates in history. Fortunately, you are able to borrow a petrol generator to avert a disastrous loss.

Lucky pants: Paisley
Lucky clip: Paper

Libra

This week sees an unwelcome return of the recurrent and rather unsavoury dreams about Shari Lewis and Lamb Chop at the local massage parlour. Venus goes direct on Thursday, which means that you will receive mixed news about your bid to become the Deputy Prime Minister's stunt-double.

Lucky lubricant: Mint jelly
Lucky wipes: Facial

Scorpio

Venus enters your fourth house this week and with it comes a feeling that something momentous is about to happen. This is however entirely misplaced as you will have one of the dullest weeks you can remember. Even a careless road crossing in front of a fire-engine leaves you uninjured. On Wednesday you will put too much butter on your toast and feel a bit queasy all morning.

Lucky colour: Beige
Lucky socks: Grey cotton

Sagittarius

This week may start with events that seem discouraging, but in fact time is on your side. Whatever setbacks you experience in the short-term, your ruler Jupiter is all set to bring a considerable boost to your fortunes. On Saturday, you will win the £15 top prize at the local Scout Troup raffle.

Lucky ticker: Yellow
Lucky number: 44

Capricorn

An unusual trine between Mars and Uranus probably means that your long-term plan to form a Rubettes tribute act with the other members of the bowls club are likely to be thwarted once more when you are again found to have a boy-soprano hidden in your briefcase. Luckily he will have the presence of mind to claim to have been playing hide-and-seek.

Lucky sentence: Suspended
Lucky destination: Coventry

Aquarius

On Saturday an interesting aspect to your ruler, Uranus, will throw everything into the melting pot. Fortunately, your partner will have the speed of thought to pull most of it out with the barbecue tongs before there is too much damage. However, you will need to apply for a new HGV License and buy a replacement ironing board. Try to be more careful in the future.

Lucky call: Lineout
Lucky Key: Robert

Pisces

As a water sign, you should be particularly careful when shopping this week as Neptune goes retrograde. This indicates that you are likely to be injured by a tin of tuna in brine that was canned rather too hastily and which survives in the salt-water in the can long enough to make a break for freedom. On Thursday, the staff canteen runs out of Danish pastries.

Lucky alternative: Belgian bun
Lucky accessory: Catcher's mitt

Weekly Forecast for 26th December to 1st January

Aries

Mars forms an unusual square with Saturn this week. You should try to avoid any game with young children that involves 'washable' felt pens, as on Tuesday the 'funny purple eyelashes' may be harder to remove than the label might suggest. The repeated application of cleansers will leave you looking somewhere between Alice Cooper and The Singing Detective. You have to choose between taking time off or fronting it out.

Lucky outfit: Black leather catsuit
Lucky boots: Silver platform

Taurus

Your irritable mood looks set to continue until Venus goes retrograde on Thursday and good fortune returns once more. On Friday, a chance meeting with a professional Mervyn King look-alike leads to investigations, which reveal that you are the legitimate heir to the Romanov crown.

Lucky egg: Fabergé
Lucky bearing: Regal

Gemini

If you are in business it's a good time. Financially you should be able to make good progress. Saturday may be particularly good, with some unexpected developments and the opportunity to have your say. The end of the week is likely to see you much in demand socially, particularly since your party-piece at the

Masonic hall which put an entirely new perspective on the old music-hall favourite of plate-spinning.

Lucky tern: Arctic
Lucky bat: Table tennis

Cancer

You're entering into a time of opportunity. Recently there have been people in your life who are just plain annoying. Yet with any luck you have avoided retaliation and have been concentrating on getting on with your own affairs. The downside to this is that at least three of them will be discovered by the end of the week and you will subsequently be cited as a co-respondent in the divorce proceedings that follow.

Lucky reputation: Tarnished
Lucky disguise: Prince Michael of Kent

Leo

You have always enjoyed being the centre of attention, and the intervention of Chiron in your sign means you will be getting more than you had anticipated as on Tuesday you learn that, after all these years, Interpol are finally onto you. On Friday you will be offered the chance to invest in a poison franchise.

Lucky sensation: Pins and needles
Lucky precaution: Antidote

Virgo

If you are studying in any way, or involved in education as teacher or pupil, you should have a very good week. Your ideas will almost certainly get off the ground although the one about the vertical-take-off and landing tuck-trolley may cause one or two raised eyebrows at the next Parent-Teacher Association meeting, as the £20,000 cost of the Pratt & Whitney Jet Engine was not in this year's budget.

Lucky excuse: A big boy did it and ran away
Lucky tent: Marquee

Libra

Many people with Sun in Libra, or Libra rising in their chart, have some important issue they have been avoiding for a while. This is the week to see to it. If it is a health matter it's not worth putting it off any longer. Regardless of what Bill at the swimming club told you, it is not normal to have three sets of eyebrows and the sooner you get something done about it the better.

Lucky band: Elastic
Lucky puppet: Troy Tempest

Scorpio

Some people with Scorpio strong in their charts will find this week brings an opportunity they've long been waiting for. If you've had an interest that you have wanted to pursue for some time, the New Moon indicates that this could be the time to do it. It will be nice and dark so even if someone does see you standing by their washing line with your pockets full, you will not be recognised.

Lucky Mac: Dirty
Lucky time: 2.20am

Sagittarius

You remain on top form. You are so full of energy that you may expect everyone else to keep up with the tough pace you are setting, but barely even give them the chance before you step in and do it for them. The gardener is likely to be philosophical about it, but your dentist will have entirely different views.

Lucky rake: Spring-time
Lucky extraction: Rolled-up copy of *Country Life*

Capricorn

This coming weekend it would be wise to keep a guard on your tongue. You know how you can mean well, but a chance remark creates untold repercussions? Well, this weekend is a prime example. You are in no position to declare war on North Korea and your feeble attempt to mobilise the local Sea Scouts is greeted with bemusement at first, followed swiftly by a call to the authorities.

Lucky window: Sash
Lucky blockage: Embolism

Aquarius

This week you can do no wrong. The sun shines, if not out of, then from the near vicinity of your ruling planet Uranus. On a visit to the wet-fish counter in Morrisons on Thursday, you will strike up a conversation with a woman during which you learn enough to be both surprised and delighted with the range and flexibility of frozen Hake loins.

Lucky sauce: Parsley
Lucky bag: Bio-degradable

Pisces

A difficult week during which Mercury goes retrograde, leading to serious misunderstandings. A casual remark over a few drinks to a rare-cattle breeds farmer about your enthusiasm for seeing running before the bulls in Pamplona is likely to have repercussions when 200 fighting bulls are released from their corral in John Lewis down the main shopping concourse of Lakeside during the January sales.

Lucky position: 1st floor balcony

Lucky retreat: Hasty

Acknowledgements

My thanks go to my Agent, Darin Jewell, of Inspira Group, Tom Chalmers of Legend Press and of course to the world's favourite Irishman, Sir Terry Wogan – who you will be delighted to hear is every bit as nice as you think he might be.

www.legendpress.co.uk

www.twitter.com/legend_press